煮人必學家常菜

Must-learn home-style recipes

目錄

Contents

煎烤
Frying & Grilling

蒸
Steaming

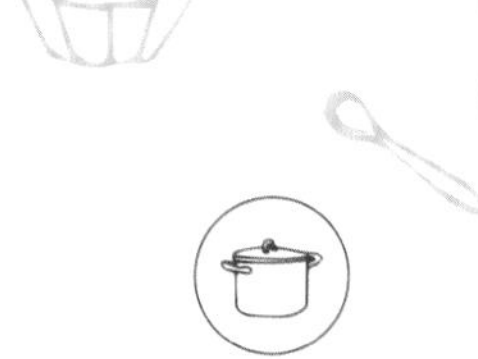

煮人的驕傲

A chef's pride

許多新手入廚的煮人，他們的熱誠可嘉，但烹調技巧不足，水準不穩定，如煮出來的菜式不美味，菜式難免受到同枱食客的冷暴力對待，煮人難免信心受挫！想煮得美味，當中有許多烹調的竅門，例如要用細火爆炒醬汁，避免醬汁有焦燶味；黃豆要浸泡數小時才煮，才能加快將豆煮至軟綿等等。在《煮人必學家常菜》內，每一個食譜都有「零失敗技巧」，讓你成功煮出美味菜式。

書內將易煮又美味的必學菜式分為燜煮、炒、蒸、煎烤等，有海鮮、魚、肉和蔬菜，這都是尋常的食材，學懂了，就有三十多道菜式和四個醬汁傍身，只要懂得融會貫通，就可變化出許多道菜式了。

現在就翻翻食譜，看你想煮哪個菜式吧！祝你樂在廚中，煮出讓你驕傲的菜式。

Novice chefs are passionate about cooking. But they are tied down by their lack of skills and experiences so that the food they make don't always hit the mark. When their hard work turns into something not particularly delicious, it would silently slip into deep oblivion on the dining table, the ultimate passive violence that frustrates the chef in the most heart-wrenching ways. To make tasty food, there are many secret tricks. You must cook the sauce over low heat for prolonged period to build depth of flavours without burning the sauce. You must soak soybeans in water for a few hours in advance before braising them to turn mushy and creamy quickly. In this cookbook, every recipe comes with a fail-proof trick that guarantees success.

The easy and delicious recipes in this book involve a range of cooking techniques, such as braising, stir-frying, steaming, frying and grilling. The most common types of ingredients and produce, such as seafood, fish, meat and veggies are also covered. When you master all the 30+ recipes and the four sauces therein, you can spin off countless dishes by applying the same logic and techniques.
Flip through this book already and find the long-overdue recipe to try out tonight. Wish you joy and fun in the kitchen.

味噌番茄醬

Miso Tomato Sauce

味道
- 味道香醇濃厚
- 帶微酸

配搭
燜煮肉類，如腩肉、一字排或豬蹄肉

成功要訣
味噌不宜久煮，宜最後加入拌勻，保留味噌香氣。

食用期限
放於雪櫃可儲存 3 日。

材料
日本味噌 2 湯匙
番茄汁 4 湯匙
紹酒 3 湯匙
冰糖 1 湯匙（舂碎）

做法
將所有材料用小火煮滾至糖溶化，備用。

Taste
- rich and complex
- mildly sour

Combinations
stewed or simmered meat, such as pork belly, spareribs or pork shoulder

Secret tricks
Miso should not be cooked for too long. Just stir it into the sauce at last to retain its fragrance.

Shelf life
It lasts in the fridge for 3 days.

Ingredients
2 tbsp miso
4 tbsp ketchup
3 tbsp Shaoxing wine
1 tbsp rock sugar, crushed

Method
Cook the ingredients over low heat until it boils and the sugar dissolves. Set aside.

蝦醬

Fermented Shrimp Paste

味道
- 鹹味適中
- 帶果皮的微香

配搭
蒸、焗、煮、炒皆宜

成功要訣
陳皮用水浸軟，刮去瓤，以免醬汁帶苦澀味。若醬汁太濃稠，最後可加油拌稀。

食用期限
放於雪櫃可儲存 2 個月。

材料
蝦膏 2 兩
陳皮茸 1 茶匙
米酒 4 湯匙
糖 1 湯匙
油 4 湯匙
熱水 4 湯匙

做法
1. 蝦膏注入熱水，壓碎拌勻，備用。
2. 全部材料拌勻，放入小鍋內以中慢火煮至散發香味即成，待涼，入瓶儲存。

Taste

- less salty taste
- with a hint of citrus warmth of the dried tangerine peel

Combinations

used in steamed, stewed, simmered and stir-fried dishes

Secret tricks

Soak the dried tangerine peel in water until soft and then scrape off its pith. Otherwise, the shrimp paste may taste bitter. If it looks too thick, thin it out with some oil at last.

Shelf life

It lasts in the fridge for up to 2 months.

Ingredients

75 g dried shrimp paste
1 tbsp finely chopped dried tangerine peel
4 tbsp rice wine
1 tbsp sugar
4 tbsp oil
4 tbsp hot water

Method

1. Put the dried shrimp paste into hot water. Crush it with a spoon and stir until well incorporated.
2. Mix all ingredients in a small pot. Heat it up over medium-low heat until fragrant. Leave it to cool. Store in sterilized bottles.

香菇肉醬

Mushroom Meat Sauce

味道

- 帶肉香及冬菇香氣
- 醬汁味香、濃郁

配搭

- 炒煮小菜
- 拌米粉、麵或飯

成功要訣

選 7 分瘦 3 分肥的豬瘺肉製成免治豬肉，令醬料香濃、有嚼口。所有材料必須炒透，令油脂分泌均勻，香口惹味。

食用期限

放於雪櫃可儲存 3 天。

做法

1. 冬菇去蒂、洗淨，用水浸軟，擠乾水分，切碎。
2. 燒熱鑊下油 3 湯匙，下乾葱茸、薑茸及蒜茸炒香，加入免治豬肉及冬菇粒炒香，濳酒，下調味料及水煮滾，轉小火煮 20 分鐘，下芡汁煮滾即成。

材料

冬菇 1 兩
免治豬肉半斤
乾葱茸 4 湯匙
薑茸 1 湯匙
蒜茸 1 湯匙
米酒 1 湯匙
水半杯

調味料

老抽、麻油各 2 湯匙
鹽 2 茶匙
糖 1 茶匙
胡椒粉少許

芡汁（拌勻）

粟粉 2 茶匙
水 3 湯匙

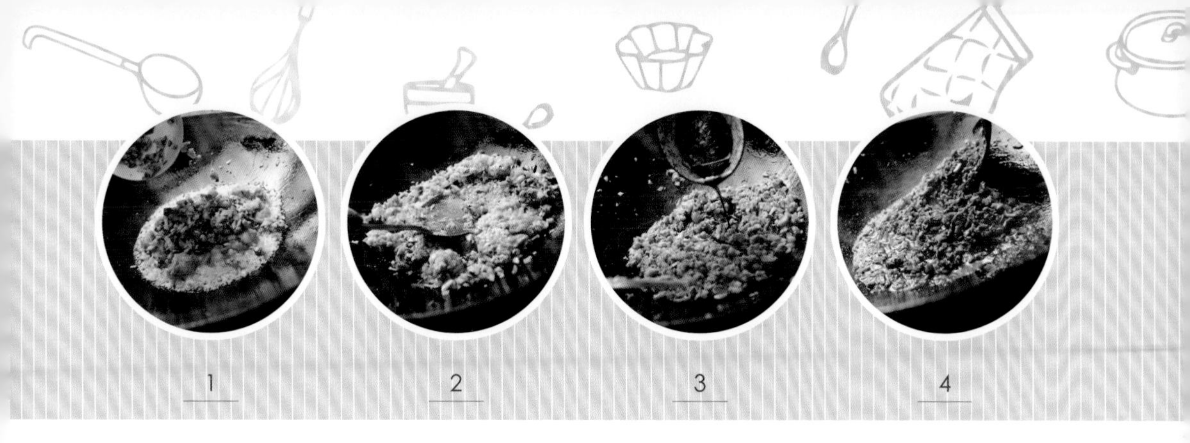

Taste

- rich meaty and mushroom flavour
- succulent and aromatic

Combinations

- stir-fries
- blanched noodles and rice vermicelli, or steamed rice

Secret tricks

Use pork shoulder cut with about 30% fat. Grind it or chop it finely for this sauce. This cut makes the sauce rich and meaty while giving a chewy texture. All ingredients must be stir-fried thoroughly so that the fat comes out into the sauce for extra fragrance.

Shelf life

It lasts in the fridge for 3 days.

Ingredients

38 g dried shiitake mushrooms
300 g ground pork
4 tbsp finely chopped shallot
1 tbsp grated ginger
1 tbsp grated garlic
1 tbsp rice wine
1/2 cup water

Seasoning

2 tbsp dark soy sauce
2 tbsp sesame oil
2 tsp salt
1 tsp sugar
ground white pepper

Thickening glaze (mixed well)

2 tsp cornflour
3 tbsp water

Method

1. Cut the stems off the mushrooms. Rinse well. Soak them in water until soft. Squeeze dry. Chop them finely.
2. Heat a wok and add 3 tbsp of oil. Stir fry shallot, ginger and garlic until fragrant. Put in the ground pork and shiitake mushrooms. Stir fry until fragrant. Press the pork with a spatula to break it into bits. Sizzle with wine. Add seasoning and water. Bring to the boil. Turn to low heat and simmer for 20 minutes. Stir in thickening glaze. Bring to the boil again.

蒜香豆豉辣椒醬

Black Bean and Garlic Chilli Sauce

味道
- 濃濃的蒜香味
- 辣味香濃

配搭
- 炒煮粉麵
- 火鍋蘸汁
- 肉類及海鮮

成功要訣
加入蒜茸及豆豉茸後，炒時必須使用小火，切勿炒焦，影響味道。

食用期限
放於雪櫃可儲存半年。

材料
指天椒 3 兩
蒜茸 2 湯匙
豆豉 2 湯匙
蝦米 2 湯匙
粟米油 1 1/2 杯

調味料
老抽 1 湯匙
鹽 1 茶匙
糖 1 1/2 茶匙

做法
1. 指天椒洗淨，去蒂，切碎；蝦米洗淨，切碎；豆豉用水沖洗，切碎（或可用石舂將材料搗爛）。
2. 下油半杯燒熱，下蝦米及指天椒炒香，加入蒜茸及豆豉茸不斷炒香，下調味料及餘下之粟米油，用小火煮至滾，再煮片刻，待涼，入瓶儲存。

Taste

- strong garlicky flavour
- hot and spicy
-

Combinations

- stir-fried noodles
- as a dip for hotpot
- meat and seafood

Secret tricks

Stir fry the mixture over low heat after adding grated garlic and crushed fermented black beans. Otherwise it might get burnt and turns bitter.

Shelf life

It lasts well in the fridge for 6 months.

Ingredients

113 g bird's eye chillies
2 tbsp grated garlic
2 tbsp fermented black beans
2 tbsp dried shrimps
1 1/2 cups corn oil

Seasoning

1 tbsp dark soy sauce
1 tsp salt
1 1/2 tsp sugar

Method

1. Rinse the bird's eye chillies. Remove the stems. Finely chop them. Set aside. Rinse the dried shrimps. Chop them. Set aside. Rinse the fermented black beans. Chop or crush them. (Or you may crush ingredients separately with a mortar and pestle.)
2. Heat a wok and pour in 1/2 cup of oil. Stir fry dried shrimps and bird's eye chillies until fragrant. Add garlic and fermented black bean. Stir continuously until fragrant. Add seasoning and the remaining corn oil. Bring to the boil over low heat. Cook briefly. Leave it to cool. Transfer into sterilized bottles.

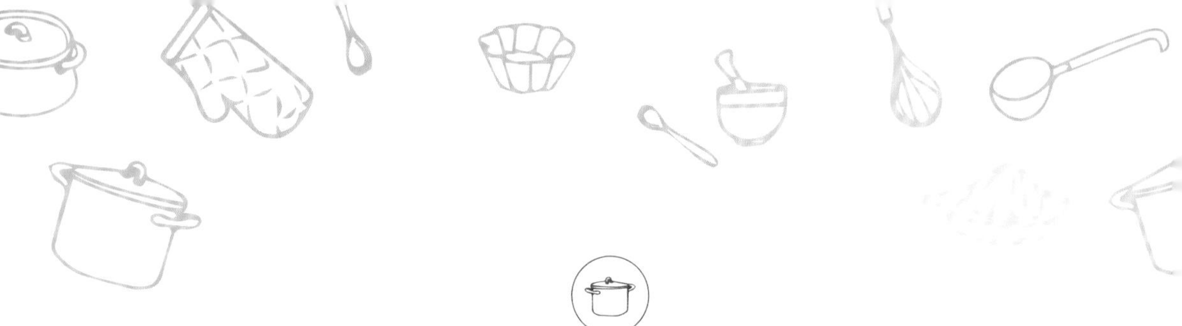

香菇肉醬炒蟶子

Stir-fried Razor Clams in Mushroom Meat Sauce

材料（4 人份量）

新鮮蟶子 1 斤
香菇肉醬 1/3 碗（做法看 P.8）
蒜茸 2 茶匙
紹酒 1 湯匙

做法

1. 蟶子用水洗淨，飛水，隔去水分。
2. 燒熱鑊下油 1 湯匙，下蒜茸炒香，加入香菇肉醬及蟶子炒勻，灒酒，炒勻，注入熱水 3 湯匙，加蓋焗 3 分鐘拌勻，上碟。

Ingredients (Serves 4)

600 g live razor clams
1/3 bowl mushroom meat sauce, refer to P.8
2 tsp grated garlic
1 tbsp Shaoxing wine

Method

1. Rinse the razor clams well. Blanch in boiling water. Drain well.
2. Heat a wok and add 1 tbsp of oil. Stir fry garlic until fragrant. Put in the mushroom meat sauce and razor clams. Stir well. Sizzle with wine. Pour in 3 tbsp of hot water. Cover the lid and cook for 3 minutes. Stir again. Save on a serving plate.

零失敗技巧
Successfully Cooking Skills

蟶子容易熟透嗎？
用水焗煮約 3 分鐘即可，太久肉質會變韌。

Do razor clams cook quickly?
They are done after cooking for 3 minutes. Do not overcook. Or else they turn tough.

蟶子如何刷洗？
用百潔布刷淨外殼，置於水喉下沖淨即可。

How do you clean the razor clams?
Scrub their shells with a scouring pad. Then rinse them well under a running tap.

瀨尿蝦肉炒韭菜花

Stir-Fried Mantis Shrimp Meat with Flowering Chinese Chives

材料（4 人份量）
瀨尿蝦（大）半斤
韭菜花 8 兩
蒜肉 4 粒（拍鬆）
紹酒半湯匙

醃料
胡椒粉少許
粟粉 1 茶匙

調味料
鹽 1 茶匙
水 2 湯匙

做法
1. 用剪刀去掉蝦頭，沿瀨尿蝦殼四周剪一圈，撕開蝦殼，取出蝦肉。
2. 用廚房紙抹乾瀨尿蝦肉，下醃料拌勻，醃 15 分鐘。
3. 韭菜花洗淨，切段。
4. 燒熱鑊下油 2 湯匙，下蒜肉爆香，加入瀨尿蝦肉炒勻，灒酒，下韭菜花及調味料炒片刻，至蝦肉全熟即成。

Ingredients (Serves 4)
300 g large mantis shrimps
300 g flowering Chinese chives
4 cloves skinned garlic, crushed
1/2 tbsp Shaoxing wine

Marinade
ground white pepper
1 tsp cornflour

Seasoning
1 tsp salt
2 tbsp water

Method
1. Cut off the shrimp heads with a pair of scissors. Cut along the edges of the shell. Tear open the shell. Remove the meat.
2. Wipe the shrimp meat with kitchen paper. Mix in the marinade and rest for 15 minutes.
3. Rinse the flowering Chinese chives, and cut into sections.
4. Heat up a wok. Add 2 tbsp of oil. Stir-fry the garlic until fragrant. Put in the shrimp meat and stir-fry well. Sprinkle with the Shaoxing wine. Add the flowering Chinese chives and seasoning. Stir-fry for a while until the shrimp meat is cooked through. Serve.

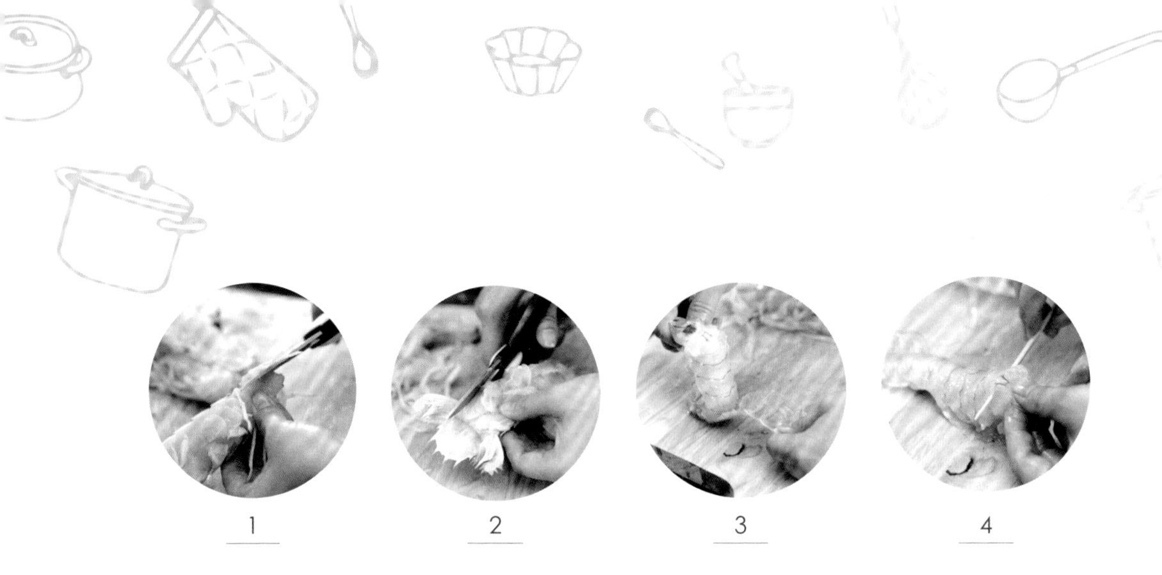

1

2

3

4

零失敗技巧
Successfully Cooking Skills

如何輕易地拆出瀨尿蝦肉？

剪去蝦頭，沿瀨尿蝦殼四周剪一圈；揭開外殼，用牙籤挑出完整蝦肉。

How to remove the meat of mantis shrimp easily?

Cut off the head with a pair of scissors. Cut along the edges of the shell. Open the shell. Take out the whole meat with a toothpick.

街市沒有大瀨尿蝦出售，可以選購細小的嗎？

雖然選購大瀨尿蝦可品嘗啖啖蝦肉，但小巧的瀨尿蝦也帶鮮味，只是蝦肉略少。

Large mantis shrimp is not available in the market. Can we choose small ones?

Large ones are meaty, but small ones are also fresh in taste with less meat.

為甚麼要索乾瀨尿蝦的水分才醃味？

一定要徹底索乾瀨尿蝦的水分才醃味，避免被溢出的水分沖淡蝦鮮味。

Why do you wipe the mantis shrimps completely dry before adding the marinade?

You must wipe them dry before marinating them. Otherwise, the juices and moisture that comes out of the mantis shrimps will dilute the marinade and make them less tasty.

黑白胡椒蝦煲

Black and White Peppers Prawn Casserole

材料（4 至 5 人份量）

中蝦 500 克
黑胡椒粒、白胡椒粒各 1 湯匙（舂碎）
乾葱茸及蒜茸各 2 茶匙
紅辣椒半隻（切碎）
香茅 1 枝（切碎）
生粉 1 湯匙
紹酒 2 茶匙

調味料

鹽 1/4 茶匙
魚露 2 茶匙
糖半茶匙
水 4 湯匙

做法

1. 中蝦剪去蝦鬚及蝦腳，於背部輕剔一刀，去掉蝦腸，抹乾備用。
2. 燒熱油 2 湯匙，中蝦撲上生粉，下油鑊略煎至七成熟，盛起。
3. 燒熱油 1 湯匙，下蒜茸、乾葱茸、黑白胡椒碎、香茅碎及紅辣椒爆香，待散發香氣，加入蝦炒勻，灒酒，傾入調味料拌勻，將全部材料轉放已塗油的瓦煲內，加蓋焗煮片刻，原煲上桌享用。

Ingredients (Serves 4-5)

500 g medium prawns
1 tbsp black peppercorns, crushed
1 tbsp white peppercorns, crushed
2 tsp finely chopped shallot
2 tsp finely chopped garlic
1/2 red chilli, chopped
1 stalk lemongrass, chopped
1 tbsp caltrop starch
2 tsp Shaoxing wine

Seasoning

1/4 tsp salt
2 tsp fish sauce
1/2 tsp sugar
4 tbsp water

Method

1. Cut away the tentacles and legs of the medium prawns. Cut a slit in the back. Remove the veins. Wipe dry and set aside.
2. Heat up 2 tbsp of oil. Coat the medium prawns with the caltrop starch. Slightly fry until they are 70% done. Set aside.
3. Heat up 1 tbsp of oil. Stir-fry the garlic, shallot, crushed black and white peppercorns, lemongrass and red chilli until aromatic. Add the medium prawns and stir-fry. Sprinkle with the wine. Pour in the seasoning and mix well. Transfer all the ingredients into a casserole spread with oil. Put a lid on and cook for a while. Serve with the casserole.

零失敗技巧
Successfully Cooking Skills

煎中蝦前先撲上生粉，有何作用？

生粉可緊鎖中蝦的鮮味，將蝦味保留其中，以免鮮味隨熱力揮發。

What is the purpose of coating medium prawns with caltrop starch before frying?

It can lock in the fresh prawn flavour; otherwise, the flavour will go with heat.

為何黑、白胡椒混合使用？

黑胡椒的香氣濃郁、味道辛辣；白胡椒的香味略淡，但卻帶陣陣芳香味，兩者混合使用，互補長短，令胡椒蝦煲惹味好吃！

Why use both black and white peppercorns?

Black peppercorn has an intense aroma with pungent taste while white peppercorn has a nice and mild fragrance. They complement each other to give the dish a sensational taste.

家裏沒有魚露，可以刪掉嗎？

魚露帶一股鮮香味，作為調味拌炒，能提升蝦的鮮味；若沒加入魚露，鮮味略遜。

There is no fish sauce at home. Can it be skipped?

Fish sauce is a nice flavoured seasoning in stir-fries. It will enhance the prawn flavour. If skipped, the dish is less tasty.

甜椒牛肝菌炒帶子

Stir-Fried Scallops with Bell Pepper and Porcini

材料（4 人份量）
急凍帶子 150 克
乾牛肝菌 30 克
蜜糖豆 100 克
紅、黃甜椒各半個（切件）
蒜茸、乾葱茸各 2 茶匙
薑 4 片
紹酒 1 茶匙

醃料
鹽 1/4 茶匙
胡椒粉少許

調味料
浸泡牛肝菌水 2 湯匙
鹽半茶匙
糖 1 茶匙
生抽 1 茶匙
麻油及胡椒粉各少許

粟粉水
粟粉 1 茶匙
水 1 湯匙
* 拌勻

做法
1. 帶子放於雪櫃下層自然解凍，飛水，橫切一刀（不切斷），下醃料拌勻。
2. 蜜糖豆撕去粗蒂及硬筋，飛水，過冷河，備用。
3. 牛肝菌用少許水浸數分鐘，去掉砂粒，再用水 1/4 杯浸半小時，擠乾水分，浸牛肝菌水留用。
4. 燒熱適量油，下帶子略泡，盛起。
5. 鑊內餘下少許油，下蒜茸、乾葱茸及薑片爆香，加入牛肝菌、紅黃甜椒略炒，濳酒，下調味料加蓋焗煮 3 分鐘。
6. 拌入蜜糖豆炒煮，最後下帶子炒勻，傾入適量粟粉水埋芡即可。

Ingredients (Serves 4)

150 g frozen scallops
30 g dried porcini
100 g snap peas
1/2 red bell pepper, cut into pieces
1/2 yellow bell pepper, cut into pieces
2 tsp finely chopped garlic
2 tsp finely chopped shallot
4 slices ginger
1 tsp Shaoxing wine

Marinade

1/4 tsp salt
ground white pepper

Seasoning

2 tbsp soaking porcini water
1/2 tsp salt
1 tsp sugar
1 tsp light soy sauce
sesame oil
ground white pepper

Cornflour solution

1 tsp cornflour
1 tbsp water
* mixed well

Method

1. Defrost the scallops in the lower chamber of the refrigerator. Scald. Cut horizontally in the middle (without cutting off). Mix with the marinade.
2. Tear off the hard stalks and strings of the snap peas. Scald and rinse with cold water. Set aside.
3. Soak the porcini in a little water for a couple of minutes. Remove the sand grains. Soak in 1/4 cup of water again for 1/2 hour. Squeeze water out. Reserve the porcini water.
4. Heat up some oil. Deep-fry the scallops lightly. Set aside.
5. Leave a little oil in the wok. Stir-fry the garlic, shallot and ginger until fragrant. Add the porcini, red and yellow bell peppers, and stir-fry slightly. Sprinkle with the wine. Add the seasoning. Put a lid on and cook for about 3 minutes.
6. Put in the snap peas. Stir-fry. Finally add the scallops. Stir-fry evenly. Thicken with cornflour solution. Serve.

零失敗技巧
Successfully Cooking Skills

帶子必須先泡油嗎？

帶子泡油後，以免炒煮時溢出水分。

The scallops must be slightly deep-fried first, must't they?

Yes, they must. It is to prevent water coming out when stir-frying.

炒帶子有特別技巧嗎？

炒帶子的動作要快捷，別過火，以免帶子變韌及縮小。

When you stir-fry the scallops, is there any special skill?

You need to do it swiftly and avoid overcooking them. Otherwise, the scallops will shrink substantially and become rubbery.

甚麼是牛肝菌？

牛肝菌是歐洲常見的菇菌，肉質肥厚，香氣豐足。在香港一般很難購買新鮮的牛肝菌，乾品於超市有售。

What is porcini?

It is a kind of mushrooms commonly found in Europe. It has a meaty texture and a rich fragrance. In Hong Kong, it is difficult to buy fresh porcini, but dried ones are sold at supermarkets.

浸泡牛肝菌的水有何作用？

牛肝菌濃郁的香氣留在水內，丟掉有點可惜，一併用作調味，能提升整道餸的味道。

What is the use of the water after soaking in the porcini?

The water has the intense aroma of porcini. It is a waste to discard it. Using it with the seasoning will improve the flavour of the dish.

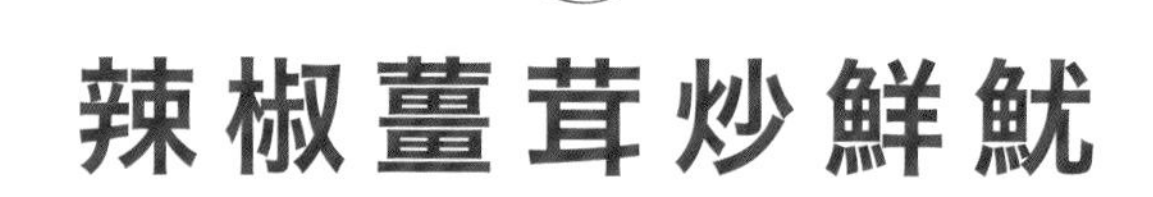

辣椒薑茸炒鮮魷

Stir-Fried Squid with Chilli and Ginger

材料（3 至 4 人份量）

新鮮魷魚 1 隻（約 12 兩）
紅椒粒半湯匙
薑茸半湯匙
葱粒 1 湯匙
紹酒半湯匙

調味料

鹽半茶匙

做法

1. 鮮魷劏好，撕去外衣，洗淨，剞成十字花狀，切塊。
2. 鮮魷放入滾水內飛水，見鮮魷收縮即盛起，隔乾水分。
3. 燒熱鑊下油 2 湯匙，下薑茸、紅椒粒及葱粒，用中大火爆香，加入鮮魷炒勻，灒酒炒片刻，灑入鹽調味，拌勻即成。

Ingredients (Serves 3-4)

1 fresh squid, about 450 g
1/2 tbsp diced red chilli
1/2 tbsp finely chopped ginger
1 tbsp diced spring onion
1/2 tbsp Shaoxing wine

Seasoning

1/2 tsp salt

Method

1. Gut the squid. Tear off the skin. Rinse and make shallow crisscross cuts with a knife. Cut into pieces.
2. Blanch the squid. When it shrinks, dish up immediately. Drain.
3. Heat up a wok. Put in 2 tbsp of oil. Stir-fry the ginger, red chilli and spring onion over medium-high heat until scented. Add the squid and stir-fry well. Sprinkle with the Shaoxing wine and stir-fry for a while. Season with the salt. Mix well. Serve.

零失敗技巧
Successfully Cooking Skills

鮮魷為何剔成十字花狀？

在鮮魷厚肉的部份剔成十字紋，令肉質容易受熱，爽滑嫩口，而且賣相吸引。

Why make shallow crisscross cuts on the squid?

Score the thick meat to make it heated up quickly. The squid is crunchy, soft and gentle. It also looks gorgeous!

鮮魷為甚麼要先飛水？

鮮魷先飛水，可減低炒煮時間，吃到魷魚之鮮香，而且肉質滑嫩。

Why do you blanch the squid in boiling water first?

Squid tends to overcook easily. It is parboiled first to partly cook it so that it takes less time to stir-fry it. You'd get to taste the umami of the squid while it is till soft and succulent.

如何炒得更惹味好吃？

增加薑茸、葱粒及紅椒的份量，用大火爆炒，倍添香氣！

How to make the flavour sensational?

Add more ginger, spring onion and red chilli and stir-fry them over high heat. It lends the dish rich fragrance!

如何挑選鮮魷？

新鮮的魷魚外形完整、身體帶光澤、魷魚尾鰭及鬚頭呈深色。

How to select squid?

Fresh squid is glossy and intact in shape with dark caudal fin and tentacles.

豆醬焗肉蟹

Mud Crabs with Puning Bean Sauce

材料（4 人份量）

肉蟹 2 隻（約 1 斤 4 兩）
潮州普寧豆醬 1 湯匙
薑 8 片
葱 6 條（切段）
粟粉 2 茶匙
紹酒適量

調味料

糖半茶匙，水半杯

做法

1. 肉蟹劏淨，處理妥當，洗淨，斬件；蟹鉗用刀拍裂，下粟粉拌勻。
2. 燒熱鑊，下油 4 湯匙，放入蟹件煎封，盛起。
3. 原鑊下薑片及半份葱段炒香，蟹件回鑊，濽酒拌勻，加入豆醬及調味料，加蓋，用中火焗煮 5 分鐘，最後放入餘下葱段拌勻即成。

Ingredients (Serves 4)

2 mud crabs (about 750 g)
1 tbsp Chaozhou Puning bean sauce
8 slices ginger
6 sprigs spring onion, sectioned
2 tsp cornflour
Shaoxing wine

Seasoning

1/2 tsp sugar
1/2 cup water

Method

1. Remove the gills and guts of the crabs. Rinse and chop into pieces. Pound the claws with a knife. Mix well with the cornflour.
2. Heat a wok. Pour in 4 tbsp of oil. Fry the crabs. Set aside.
3. Stir-fry the ginger and half of the spring onions in the same wok until fragrant. Put in the crabs. Sprinkle with the wine and stir well. Add the bean sauce and the seasoning. Cover with a lid and cook over medium heat for 5 minutes. Stir in the rest of the spring onions at last. Serve.

零失敗技巧
Successfully Cooking Skills

為何使用大量薑葱？

薑葱散發獨特香氣，是起鑊炒蟹之最佳配料，令蟹件滲滿惹味的薑葱香味。

Why use a great amount of ginger and spring onion?

With unique fragrance, they are the best ingredients for stir-frying crabs which give the appetizing smell of ginger and spring onion.

蟹件需要煎封多久？

見蟹件轉成紅色即可，毋須完全熟透，因隨後需經焗煮步驟。

How long should we fry the crabs?

Fry until they turn red. It is not necessary to make them fully done because they will be cooked again later.

買來一大瓶普寧豆醬，還可用來弄甚麼菜式？

炒油麥菜、炒通菜、蒸魚、蒸排骨、焗雞件、蘿蔔煮魚等…美味無窮！

What kind of dishes can be made so as to fully use the big bottle of Puning bean sauce?

You can use it to stir-fry Indian lettuce or water spinach; to steam fish or pork spareribs; to cook chicken; or to braise radish with fish...All are very delicious.

蟹鉗為何用刀拍裂？

令香氣滲入蟹鉗內，而且食用時容易處理。

Why pound the crab claws with a knife?

To let the aroma of the ingredients permeate the cracked claws. It also helps shell the claws easily when eating.

蝦醬炒魷魚筒

Stir-fried Baby Squids in Fermented Shrimp Paste

材料（4 人份量）

新鮮小魷魚筒 12 兩
西芹 2 枝
蒜肉 3 粒
蝦醬 1 茶匙（做法看 P.6）
糖半茶匙
紹酒半湯匙

芡汁（拌勻）

粟粉 1 茶匙，水 2 湯匙

做法

1. 西芹撕去老筋，洗淨，切段，飛水備用。
2. 魷魚筒抽出鬚頭及內臟，去掉軟骨，洗淨。
3. 燒熱鑊下油 2 湯匙，下蒜肉及蝦醬炒香，下魷魚筒炒勻，灒酒，炒勻，加入西芹及糖炒片刻，最後下芡汁埋芡即成。

零失敗技巧
Successfully Cooking Skills

如何撕去西芹老筋？
用手撕出粗絲，將粗纖維拉盡，或用刀削去表皮的纖維。

How do you tear off the tough veins on celery?
Just snap it with your hand and pull off the veins. Or use a peeler to peel off some of the fibres on the skin.

可以原隻魷魚炒嗎？
可以，更可吸收墨汁的營養，做法相同。

Can I stir fry the baby squids in whole?
Yes, you can. You actually get to eat the ink, which is nutritious. In that case, you can skip the dressing step of the squids and cook them in whole. The rest of the method will be the same.

Ingredients (Serves 4)

450 g fresh baby squids
2 celery stems
3 cloves garlic
1 tsp homemade fermented shrimp paste, refer to P.6
1/2 tsp sugar
1/2 tbsp Shaoxing wine

Thickening glaze (mixed well)

1 tsp cornflour
2 tbsp water

Method

1. Tear off the tough veins on the celery. Rinse and cut into short lengths. Blanch in boiling water. Drain.
2. Dress the squids by pulling out their tentacles, head and innards. Remove the bone. Rinse well.
3. Heat a wok and add 2 tbsp of oil. Stir fry garlic cloves and fermented shrimp paste until fragrant. Put in the squids and stir well. Sizzle with wine. Stir further. Add celery and sugar. Mix well. Stir in the thickening glaze and toss well. Serve.

山楂咕嚕肉

Sweet and Sour Pork in Hawthorn Sauce

材料（4 人份量）

急凍梅頭肉 200 克
洋葱半個（切件）
青甜椒半個（去籽、切件）
大紅辣椒半隻（去籽、切件）
菠蘿 2 片（切件）
蒜肉 1 粒（剁碎）
蛋汁 1 湯匙
生粉 4 湯匙

醃料

鹽及糖各半茶匙
生抽、紹酒及生粉各 1 茶匙
胡椒粉少許

汁料

山楂 40 克
山楂餅 2 小筒
酸梅 2 粒
片糖碎 2 湯匙
白醋 3 湯匙
茄汁 3 湯匙
鹽 1/4 茶匙

芡汁

生粉 2 茶匙
水 1 湯匙
＊ 拌勻

做法

1. 急凍梅頭肉放於雪櫃下層自然解凍，洗淨，切厚件，用刀尖在肉上輕剔數刀，再用刀背拍鬆，下醃料略醃。
2. 將汁料的山楂洗淨，用 3/4 杯水煮約 20 分鐘，加入山楂餅、酸梅肉、鹽及片糖煮約 10 分鐘，過濾，再與白醋及茄汁煮至稠，備用。
3. 梅頭肉及蛋汁拌勻，蘸上生粉。
4. 燒熱適量油，下梅頭肉炸至金黃色，盛起，待片刻，再炸一遍。
5. 燒熱少許油，下青甜椒略炒，盛起。
6. 將洋葱、蒜茸及紅辣椒再炒，傾入山楂汁及菠蘿煮片刻，加入適量生粉水埋芡，最後拌入梅頭肉及青椒即成。

零失敗技巧
Successfully Cooking Skills

為甚麼要在梅頭肉輕剔數刀，並用刀背拍鬆？
因為可以令肉質軟腍，而且調味醬汁容易滲進肉內。

我弄的咕嚕肉，很難做到內脆的口感，有何秘訣？
最重要是梅頭肉最後才與汁料混和，輕輕拌勻即可，而且必須盡快享用。
It is difficult for me to make sweet and sour pork crispy. What is the secret?
It is most important to mix the pork collar-butt with the sauce at last. Just gently mix them and serve immediately.

加添了山楂及山楂餅，食味有何不同？
醬汁甜中帶微酸，開胃醒胃，還有山楂淡淡的清香味，伴肉享用，滋味無窮！
What is the difference in flavour by adding hawthorn and haw flakes?
It gives the sweet sauce a light sour taste which is appetizing. Hawthorn has a nice aroma which adds flavour to the meat. It is palatable!

梅頭肉必須與蛋汁輕拌嗎？
能保留梅頭肉的肉汁，肉質軟滑，緊記下鍋炸前才拌入蛋汁。
Is it necessary to mix the pork collar-butt with egg wash?
It can keep the meat juicy and tender. Remember to stir in the egg wash right before deep-frying.

Ingredients (Serves 4)

200 g frozen pork collar-butt
1/2 onion, cut into pieces
1/2 green bell pepper, deseeded; cut into pieces
1/2 red pepper, deseeded; cut into pieces
2 slices pineapple, cut into pieces
1 clove skinned garlic, finely chopped
1 tbsp egg wash
4 tbsp caltrop starch

Marinade

1/2 tsp salt
1/2 tsp sugar
1 tsp light soy sauce
1 tsp Shaoxing wine
1 tsp caltrop starch
ground white pepper

Sauce

40 g dried hawthorn
2 small rolls haw flakes
2 pickled plums
2 tbsp crushed rock sugar
3 tbsp white vinegar
3 tbsp ketchup
1/4 tsp salt

Thickening glaze

2 tsp caltrop starch
1 tbsp water
* mixed well

Method

1. Defrost the pork collar-butt in the lower chamber of the refrigerator. Rinse and cut into chunks. Make a few cuts on the meat with the tip of a knife. Tenderize by pounding with the back of the knife. Mix with the marinade. Rest for a while.
2. Rinse the hawthorn. Cook with 3/4 cup of water for about 20 minutes. Add the haw flakes, pickled plum flesh, salt and rock sugar. Cook for about 10 minutes. Filter the sauce. Cook with the white vinegar and ketchup until the sauce reduces. Set aside.
3. Mix the pork collar-butt with the egg wash. Coat with the caltrop starch.
4. Heat up some oil. Deep-fry the pork collar-butt until golden. Remove. Leave for a while. Deep-fry again.
5. Heat up a little oil. Roughly stir-fry the green bell pepper. Remove.
6. Stir-fry the onion, garlic and red pepper. Pour in the hawthorn sauce and pineapple. Cook for a while. Stir in the thickening glaze and toss well. Finally mix in the pork collar-butt and green bell pepper to finish.

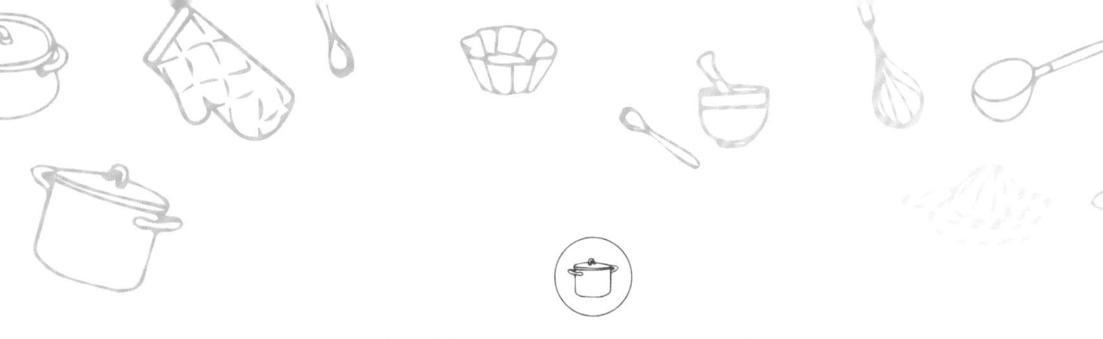

鮮菌菜苗炒雞肉

Stir-fried Chicken with Mushrooms and Baby Flowering Chinese Cabbage

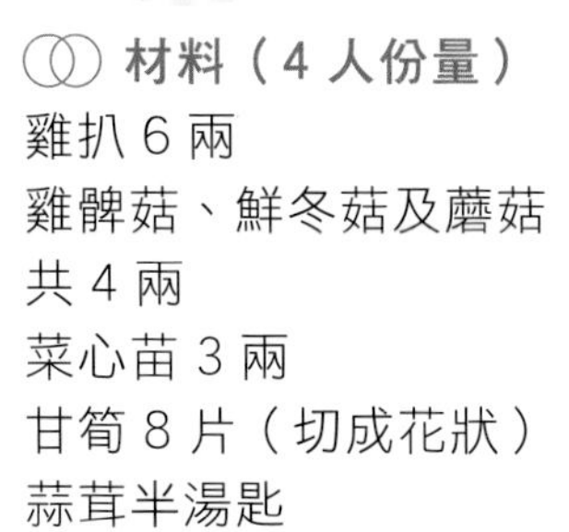

材料（4 人份量）

雞扒 6 兩
雞髀菇、鮮冬菇及蘑菇共 4 兩
菜心苗 3 兩
甘筍 8 片（切成花狀）
蒜茸半湯匙

醃料

鹽 1/3 茶匙
雞粉半茶匙
胡椒粉少許
生粉 1 茶匙
油半湯匙

調味料

鹽半茶匙
蠔油 1 茶匙
糖 1/3 茶匙
麻油少許

做法

1. 鮮冬菇及蘑菇去蒂、沖淨，切條；雞髀菇沖淨，切條；將以上雜菌用油炒至略軟，盛起備用。
2. 雞扒去皮、切條，加入醃料拌勻醃 10 分鐘。
3. 菜心苗放入油鑊內略炒，加入鹽 1/4 茶匙炒透，盛起。
4. 熱鑊下油，放入雞柳及蒜茸炒至全熟，再加入雜菌、菜心苗及調味料拌炒，最後加入甘筍花拌勻，上碟即成。

零失敗技巧
Successfully Cooking Skills

試過煮草菇，口感軟腍難吃，為甚麼？

菇菌類不宜太早洗淨，最好沖淨後 10 分鐘內烹調（或用布抹淨），以免吸入太多水分，影響口感。

I have cooked straw mushrooms before. They are too tender and distasteful. Why?

Do not rinse mushrooms too early. It is better to cook them within 10 minutes after rinsing (or clean them with a cloth); otherwise, they will absorb too much water and will not be delicious.

雞扒必須去皮嗎？

去皮可減掉油分，吃得更健康；若喜歡雞皮炒透後的香滑感覺，保留雞皮亦可，悉隨尊便。

Is it necessary to skin the chicken?

It can reduce oil and let you have a healthy diet. If you like the smooth fried skin, you may keep it.

Ingredients (Serves 4)

225 g boneless chicken thigh
150 g king oyster mushrooms, fresh black mushrooms and button mushrooms
113 g baby flowering Chinese cabbage
8 slices carrot, cut into flower shape
1/2 tbsp finely chopped garlic

Marinade

1/3 tsp salt
1/2 tsp chicken bouillon powder
ground white pepper
1 tsp caltrop starch
1/2 tbsp oil

Seasoning

1/2 tsp salt
1 tsp oyster sauce
1/3 tsp sugar
sesame oil

Method

1. Remove the stalks of the fresh black mushrooms and button mushrooms. Rinse and cut into strips. Rinse the king oyster mushrooms. Cut into strips. Stir-fried the mixed mushrooms with oil until slightly soft. Set aside.
2. Skin the chicken. Cut into strips. Mix with the marinade and rest for 10 minutes.
3. Stir-fry the baby flowering Chinese cabbage with oil roughly. Season with 1/4 tsp of salt and stir-fry thoroughly. Set aside.
4. Add oil in a heated wok. Stir-fry the chicken with garlic until fully cooked. Put in the mixed mushrooms, baby flowering Chinese cabbage and seasoning. Stir-fry. Finally mix in the carrot. Serve.

蠔油薑葱炒牛肉

Stir-Fried Beef with Ginger and Spring Onion in Oyster Sauce

材料（4 人份量）

冰鮮牛琢肉 250 克
薑 80 克（切薄片）
葱 60 克（切度，分開葱白及青葱）
蒜肉 2 粒（略拍）
紹酒 2 茶匙

醃料

生抽 1 茶匙
糖半茶匙
紹酒、粟粉及水各 1 湯匙
麻油及胡椒粉各少許
油 1 湯匙（後下）

調味料

鹽及糖各半茶匙
生抽 1 茶匙
蠔油 2 茶匙
水 2 湯匙
粟粉 1 茶匙

做法

1. 冰鮮牛肉洗淨，抹乾，放入冰格冷藏 1 小時，順橫紋切成薄片，與醃料拌勻。
2. 燒熱油 1 湯匙，下薑片、蒜肉及葱白爆香，盛起。
3. 燒熱少許油，放入牛肉平均分佈，略煎片刻，反轉後再煎，略炒推散。
4. 加入已炒香的薑片、蒜肉、葱白及青葱，灒酒，加入調味料兜勻，至汁液濃稠即可。

零失敗技巧
Successfully Cooking Skills

牛肉質地軟軟的，難以切成薄片，怎辦？
將牛肉放入冰格冷藏 1 小時，令肉質略硬，方能切出薄薄的牛肉片。
The beef is so soft that it is difficult to be sliced thinly. How to do?
Freeze the beef for 1 hour to make it a bit tough. Then you can cut it into thin slices.

我炒的牛肉又粗又韌，怎辦？
一 . 橫紋切牛肉；二 . 毋須下鹽醃，加少許水拌勻，最後下油略拌，緊封水分；三 . 牛肉別亂炒，推散兜炒。
The beef I made is thick and chewy. How to do?
1. Cut the beef across its grains. 2. To marinate, mix the beef with a little water and then oil to seal in the moisture. No salt is needed. 3. Do not stir-fry the beef roughly. Scatter the meat and stir-fry.

必須用上大量的薑葱嗎？
用大量的薑葱起鑊，是此菜之特色，炒出來的牛肉帶薑葱香味！
Is it necessary to use lots of ginger and spring onion?
Using lots of them is the characteristic of this dish. It gives the stir-fried beef the fragrance of ginger and spring onion.

Ingredients (Serves 4)

250 g chilled beef rump
80 g ginger, thinly sliced
60 g spring onion, cut into sections; white and green parts separated
2 cloves skinned garlic, slightly crushed
2 tsp Shaoxing wine

Marinade

1 tsp light soy sauce
1/2 tsp sugar
1 tbsp Shaoxing wine
1 tbsp cornflour
1 tbsp water
sesame oil
ground white pepper
1 tbsp oil (added later)

Seasoning

1/2 tsp salt
1/2 tsp sugar
1 tsp light soy sauce
2 tsp oyster sauce
2 tbsp water
1 tsp cornflour

Method

1. Rinse the beef and wipe dry. Freeze for 1 hour. Cut across the grains into thin slices. Mix with the marinade.
2. Heat up 1 tbsp of oil. Stir-fry the ginger, garlic and white part of the spring onion until scented. Set aside.
3. Heat up a little oil. Put in the beef evenly. Slightly fry for a while. Flip over and fry. Slightly stir-fry and scatter.
4. Add the ginger, garlic, white and green parts of the spring onion. Sprinkle with the wine. Finally put in the seasoning and stir-fry until the sauce reduces. Serve.

1

2

3

銀絲香芹花蛤鍋

Clams with Vermicelli and Chinese Celery in Clay Pot

材料（4 人份量）

花蛤 1 1/2 斤
乾粉絲 1 包（50 克）
芹菜 2 棵
紅辣椒 1 隻
蒜茸 1 湯匙
清雞湯 500 毫升
熱水 300 毫升

調味料

鹽適量

做法

1. 花蛤洗淨，放入清水內，下鹽 1 湯匙浸 1 小時，取出。
2. 粉絲浸軟；芹菜切段；紅辣椒切圈。
3. 砂鍋內下適量油，加入紅辣椒及蒜茸爆香，放入花蛤煮片刻，注入清雞湯及熱水，加蓋。
4. 煮至花蛤的外殼剛張開（約 3 分鐘），加入粉絲及芹菜煮滾片刻，下鹽調味，趁熱品嘗。

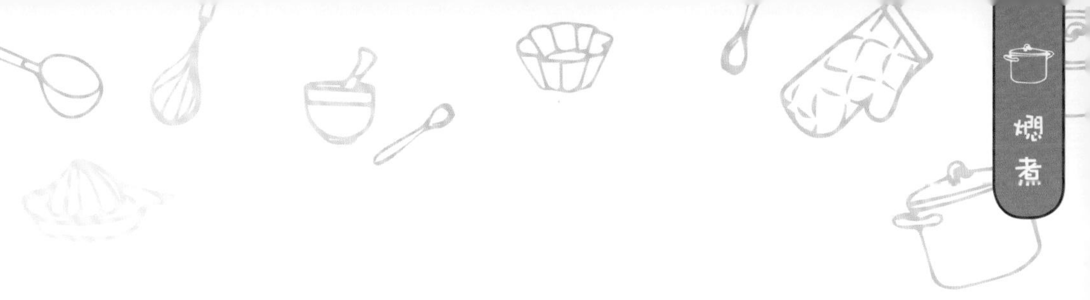

Ingredients (Serves 4)

900 g clams
1 pack (50 g) vermicelli
2 stalks Chinese celery
1 red chilli
1 tbsp minced garlic
500 ml chicken broth
300 ml hot water

Seasoning

salt

Method

1. Rinse clams. Soak in water with 1 tbsp of salt for 1 hour. Drain.
2. Soak vermicelli in water until soft. Section Chinese celery. Cut red chilli into rings.
3. Put oil into a clay pot. Add red chilli and minced garlic. Fry until fragrant. Put in clams and cook for a while. Pour in chicken broth and hot water. Cover the lid.
4. Cook for about 3 minutes until the shells of clams open. Add vermicelli and Chinese celery. Cook for a while and season with salt. Serve hot.

零失敗技巧
Successfully Cooking Skills

已用鹽水預先浸花蛤，但花蛤仍藏沙粒，怎辦？

預先用鹽水浸花蛤及洗淨後，應乾淨無沙粒。若情況無改善，建議選擇到其他海鮮供應商購買。

What can be done if clams still have sand after soaked in salt water?

Clams should be clean without sand after soaked in salt water and rinsed. If the situation cannot be improved, it is suggested to buy clams at other seafood stalls.

花蛤可浸養至翌日才煮嗎？

不可，因家裏調校的鹽水始終不及花蛤生活的大海，環境轉變了，花蛤容易死掉。

Can clams be soaked until the next day?

No. It is because salt water seasoned at home is not the same as the sea clams lived. They would die easily.

芹蒜豆腐泡煮沙鯭魚

File Fish and Beancurd Puffs with Chinese Celery and Leeks

材料（4 人份量）
沙鯭魚 1 斤
豆腐泡 4 兩
芹菜 1 棵
青蒜 2 棵
橄欖菜 1 1/4 湯匙
薑 4 片

醃料
胡椒粉少許
粟粉 1 茶匙

調味料
生抽 2 茶匙
糖半茶匙

做法
1. 沙鯭魚洗淨，抹乾水分，下醃料抹勻，備用。
2. 豆腐泡飛水，過冷河，壓乾水分。
3. 芹菜只取菜莖，洗淨，切段；青蒜切去根部，洗淨，切段。
4. 燒熱鑊下油 2 湯匙，下薑片炒香，加入沙鯭魚煎至微黃，下豆腐泡、調味料及熱水 1 1/4 杯，加蓋焗煮 5 分鐘，最後加入芹菜、青蒜及橄欖菜煮片刻即成。

Ingredients (Serves 4)
600 g file fish
150 g beancurd puffs
1 sprig Chinese celery
2 green garlic
1 1/4 tbsp olive-pickled leaf mustard
4 slices ginger

Marinade
ground white pepper
1 tsp cornflour

Seasoning
2 tsp light soy sauce
1/2 tsp sugar

Method
1. Rinse the fish and wipe dry. Add marinade and mix well. Set aside.
2. Blanch the beancurd puffs in water. Rinse with cold water. Squeeze dry.
3. Pick off the leaves of the Chinese celery and discard them. Use the stems only. Rinse well and cut into short lengths. Set aside. Cut off the roots of the green garlic. Rinse and cut into short lengths.
4. Heat a wok and add 2 tbsp of oil. Stir fry ginger until fragrant. Fry the file fish until lightly browned on both sides. Put in the beancurd puffs, seasoning and 1 1/4 cups of hot water. Cover the lid and cook for 5 minutes. Add Chinese celery, green garlic and olive-pickled leaf mustard at last. Cook briefly and serve.

零失敗技巧
Successfully Cooking Skills

為何最後加入橄欖菜？

因為久煮會減少橄欖菜的香氣。

Why do you put in the olive-pickled leaf mustard at the very last?

That would make sure the delicate fragrance of the olive-pickled leaf mustard will not be destroyed by prolonged cooking.

先煎沙鯭魚有何作用？

焗煮時，以免魚肉容易散開！

Why do you fry the file fish in oil first?

That would firm up the fish and make it less likely to fall apart in the cooking process.

甚麼是青蒜？

由獨子蒜長出的青蒜，蒜味濃郁，吸油力高，是冬季的菜蔬。

What are green garlic?

These are the plants grown from single clove garlic bulbs. Green garlic have a strong garlicky taste and pick up much oil. They are in season during the winter season.

小白菜魚餃湯鍋

Fish Dumplings and Small White Cabbages in Fish Broth

魚餃材料（4 人份量）
鯪魚肉 1 片（約 5 兩）
臘腸半條
芫茜 1 棵
葱 1 條
圓形水餃皮 10 張

配料
小白菜 4 兩

魚湯材料
鮮魚 1 斤
薑 2 片
滾水 1.25 公升

調味料（1）
鹽適量

調味料（2）
鹽 1/4 茶匙
胡椒粉少許
水及生粉各 2 茶匙

準備工夫（魚湯）

- 燒熱油，下鮮魚及薑片煎香，放入魚袋內，注入滾水煮約 1 小時，至餘下 2 1/2 杯（625 毫升）魚湯，下鹽調味。

準備工夫（魚餃）

- 用匙羹依逆紋刮出鮫魚肉。
- 臘腸隔水蒸 10 分鐘，待涼，切細粒。
- 芫茜及葱切粒。
- 魚茸、臘腸、芫茜、葱及調味料（2）放於大碗內，順一方向攪拌成魚餃餡料，分成 10 份。
- 將餡料放入水餃皮內，用水輕抹皮邊，摺成呈窩形的魚餃（看圖）。

做法

1. 煮滾一鍋水，下魚餃煮熟，盛起。
2. 取出魚湯內之魚袋，放入小白菜煮熟，下魚餃煮滾即可。

Ingredients of fish dumplings

1 piece (about 175 g) mackerel flesh
1/2 preserved sausage
1 stalk coriander
1 sprig spring onion
10 round dumpling wrappers

Ingredients of fish broth

600 g fresh fish
2 slices ginger
1.25 liters boiling water

Condiment

150 g small white cabbages

Seasoning (1)

salt

Seasoning (2)

1/4 tsp salt
ground white pepper
2 tsp water
2 tsp caltrop starch

Preparation (fish broth)

- Heat oil in wok. Fry fish and ginger slices until fragrant. Put the fish into a cloth bag and cook in boiling water for about 1 hour until 2 ½ cups (625 ml) of broth remained. Season with salt.

Preparation (fish dumplings)

- Scoop flesh from mackerel with a spoon along the vein in a reverse direction.
- Steam preserved sausage for 10 minutes and set aside to let cool. Cut into small dices.
- Chop coriander and spring onion.
- Put fish flesh, preserved sausage, coriander, spring onion and seasoning (2) into a large bowl. Stir in one direction to become the filling and divide into 10 portions.
- Put the filling on top of the wrappers. Dip water at the sides and wrap into nest-shaped fish dumplings (as shown).

Method

1. Bring a pot of water to the boil. Add the fish dumplings and cook until done. Drain.
2. Remove the cloth bag from the fish broth. Add small white cabbages and cook until done. Put in fish dumplings and bring to the boil. Serve.

1

2

3

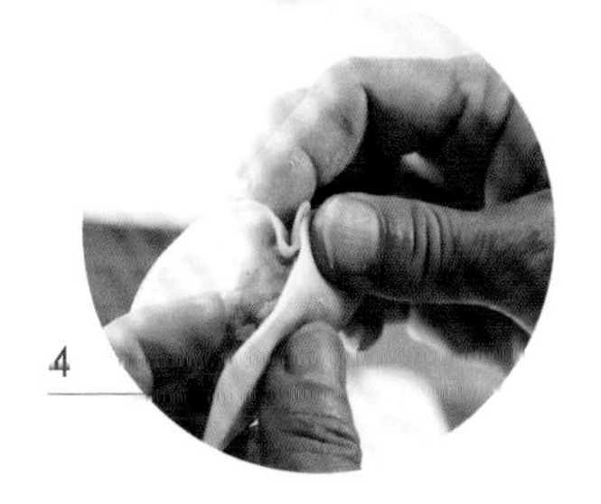
4

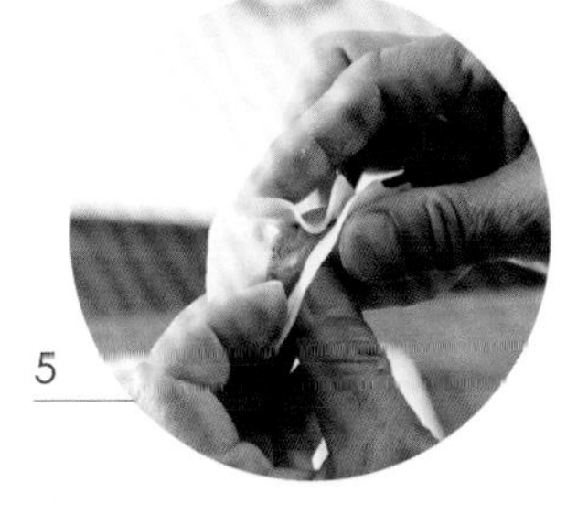
5

6

零失敗技巧
Successfully Cooking Skills

哪種魚可打成魚茸？

鹹水魚檔有售之鮫魚、九棍；淡水魚檔之鯪魚皆可。

Which kind of fish can make paste?

Mackerel and lizard fish bought from saltwater fish stalls or dace bought from freshwater fish stalls can do.

包摺魚餃有何竅門？

1. 釀入適量之餡料；2. 平均地捏摺魚餃；3. 封口時要穩固，必定能成功包成魚餃。

What are the tips of wrapping fish dumplings?

Firstly, stuff in suitable amount of fillings; secondly, knead the fish dumplings evenly; and thirdly, binding the ends well can help you make fish dumplings successfully.

通常用哪款鮮魚熬湯？

牛鰍魚、梭羅魚、三文魚骨、九棍及木棉魚皆可熬成魚湯，鮮甜美味！

What kind of fresh fish is usually used to make fish broth?

Flathead fish, solo fish, salmon bones, lizard fish or big-eyed fish can be cooked into tasty fish broth.

酒煮東風螺

Babylon Shells in Spicy Wine Sauce

材料（4 人份量）

東風螺 1 斤
紅辣椒 1 隻（開邊）
蒜肉 3 粒
九層塔 2 棵
豆豉蒜香辣椒醬 2 茶匙
（做法看 P.10）

調味料

紹酒半杯
上湯 1 量杯
糖半茶匙

做法

1. 東風螺用 1 湯匙粗鹽擦淨外殼，洗淨。
2. 燒熱鑊下油 1 湯匙，下蒜肉及辣椒醬炒香，注入調味料煮滾，加入東風螺煮約 5 至 7 分鐘（或至東風螺全熟），最後下紅辣椒及九層塔拌勻煮滾即成。

Ingredients (Serves 4)

600 g Babylon shells
1 red chilli, cut into halves
3 cloves garlic
2 sprigs Thai basil
2 tsp black bean and garlic chilli sauce, refer to P.10

Seasoning

1/2 cup Shaoxing wine
1 cup stock
1/2 tsp sugar

Method

1. Rub the shells with 1 tbsp of coarse salt. Rinse well.
2. Heat a wok and add 1 tbsp of oil. Stir fry garlic and chilli sauce until fragrant. Pour in the seasoning and bring to the boil. Put in the shells. Cook for 5 to 7 minutes (or until all shells are done). Lastly put in red chilli and Thai basil. Stir well and bring to the boil again. Serve.

零失敗技巧
Successfully Cooking Skills

如何保留酒香味？

若希望上桌時酒香撲鼻，須最後灑入紹酒 2 湯匙煮滾，關火，香氣四溢。

How do you keep the wine fragrance?

To an extra scent of wine, you might sprinkle 2 tbsp of Shaoxing wine at last. Bring to the boil and turn off the heat.

為何用粗鹽擦淨東風螺外殼？

由於東風螺長在石隙間，用粗鹽能徹底擦淨外殼，進食時更衛生。

Why do you rub the shells with coarse salt?

As the shells live in crevices, they might have slime or moss on them. Rubbing them with coarse salt makes sure they are clean.

如何得知螺肉全熟？

螺肉呈收縮狀，用竹籤容易刺入即代表螺肉已全熟。

How do you know the shells are done?

Their flesh would shrink when done. If your can pierce through its flesh easily with a bamboo skewer, it is done.

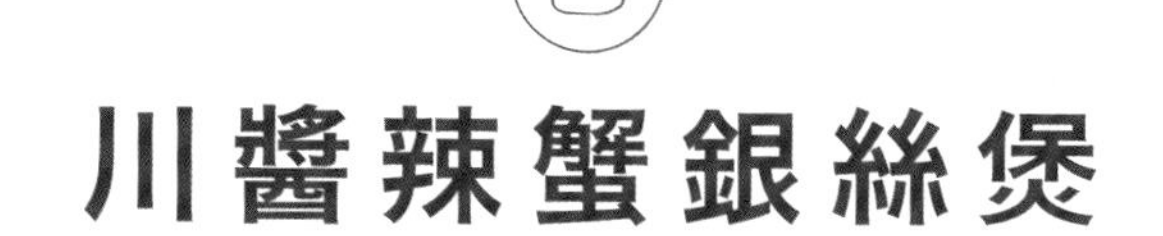

川醬辣蟹銀絲煲

Sichuan Spicy Crab and Vermicelli in Clay Pot

材料（3 至 4 人份量）

肉蟹 1 隻（約 1 斤）
粉絲 1 1/4 兩（50 克）
豆瓣醬 1 湯匙
辣椒乾 1 隻
薑 2 片
蒜茸 1 湯匙
葱 1 條
生粉適量

調味料

上湯 1 1/3 杯
鹽 1/3 茶匙

做法

1. 肉蟹洗擦乾淨，去內臟，斬件，瀝乾水分。
2. 粉絲用清水浸透，剪成段，備用。
3. 辣椒乾洗淨，浸水約 10 分鐘至軟身，切片；葱切段。
4. 蟹件灑上生粉，用適量油半煎炸至蟹件轉成紅色，盛起。
5. 燒熱瓦鍋，下油爆香豆瓣醬、辣椒乾、薑片及蒜茸，加入蟹件，注入調味料煮滾 6 分鐘，加入粉絲多煮 2 分鐘，下葱段再煮片刻，原鍋上桌品嘗。

Ingredients (Serves 3 to 4)

1 mud crab, about 600 g
50 g vermicelli
1 tbsp spicy soybean sauce
1 dried chilli
2 slices ginger
1 tbsp minced garlic
1 sprig spring onion
caltrop starch

Seasoning

1 1/3 cups broth
1/3 tsp salt

Method

1. Rub and wash mud crab. Remove the entrails and chop up into pieces. Drain.
2. Soak vermicelli in water thoroughly. Cut into sections with a pair of scissors. Set aside.
3. Rinse dried chilli. Soak in water for about 10 minutes until soft and slice. Section spring onion.
4. Sprinkle caltrop starch over the crab. Shallow-fry in oil until turns red and drain.
5. Heat clay pot and add oil. Stir-fry spicy soybean sauce, dried chilli, ginger slices and minced garlic quickly until fragrant. Add the crab and pour in seasoning. Cook for 6 minutes. Put in the vermicelli and cook for 2 minutes more. Add sectioned spring onion and cook for a while. Serve.

零失敗技巧
Successfully Cooking Skills

如何烹調得更健康？

蟹件用半煎炸之方式代替泡油，既可節省食油，又令身體更健康。

How to cook this dish in a healthy way?

Shallow-fry the crab instead of blanching it in oil not only saves oil but also healthier for the body.

為甚麼要用油爆香豆瓣醬、辣椒乾、薑片及蒜茸？

先用油爆香豆瓣醬、辣椒乾、薑片及蒜茸，可讓這些醬料、香料味道更惹味，與蟹同炒，香辣味濃，具四川辣菜之特色。

Why do you fry the spicy bean sauce, dried chilli, sliced ginger and grated garlic in oil first?

They are fried in hot oil over high heat until fragrant first before other ingredients are added. This step brings out the fragrance and aromatic oils in them. When the crab is tossed in the spice mixture, the aromas will be richer and more complex. This is a unique characteristic of Sichuan cuisine.

這道菜可以香而減辣嗎？

可以酌量剔走辣椒籽。

Can I retain the heady aroma of this dish without the fiery spiciness?

Yes. You can remove some or all of the chilli seeds to make it milder.

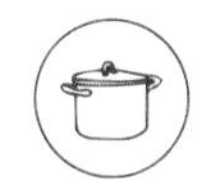

農家燜五花腩

Stewed Pork Belly

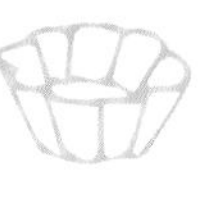

材料（4 人份量）

急凍五花腩 600 克
南乳 2 茶匙
磨豉醬 2 茶匙
陳皮 1 角（浸軟、刮瓤）
八角 2 粒
薑 6 片
蒜肉 4 粒（略拍）
紹酒 1 湯匙

調味料

鹽 1/4 茶匙
冰糖碎 2 茶匙
生抽及老抽各 2 茶匙

做法

1. 五花腩放於雪櫃下層自然解凍，放入滾水內煮約 20 分鐘，盛起，過冷河，切成厚件。
2. 燒熱少許油，下薑片及蒜肉爆香，再加入南乳及磨豉醬炒至散發香味。
3. 加入五花腩炒勻，灒紹酒，最後下調味料、水 3 杯、陳皮及八角燜煮約 1 1/2 小時，至五花腩酥腍即可。

零失敗技巧
Successfully Cooking Skills

五花腩的脂肪含量高，我怕膽固醇，怎辦？

教你一個簡單的做法：於前一天燜煮妥當，待涼後，油分浮面，放入雪櫃令油分凝固，翌日翻熱前去掉油脂，脂肪絕對下降不少，而且隔晚燜煮的五花腩，肉香入味！

Pork belly contains lots of fat. I am scared of cholesterol. How to do?

Here is a simple method: Cook the pork belly one day in advance. When it cools down, oil will surface. Refrigerate to let the oil set. Remove the fat before heating up on next day. The fat is largely reduced while the pork belly is more flavourful by resting overnight.

急凍的五花腩，肉質偏硬嗎？

絕對不會！只要你燜煮得宜，加上時間充足，五花腩會酥香綿軟。

Is the frozen pork belly tough in texture?

Absolutely not! Only by cooking properly with plenty of time, you will easily taste soft and fragrant pork belly meat.

Ingredients (Serves 4)

600 g frozen pork belly
2 tsp red fermented beancurd
2 tsp ground bean sauce
1/4 dried tangerine peel, soaked to soften; scraped off the pith
2 star aniseeds
6 slices ginger
4 cloves skinned garlic, slightly crushed
1 tbsp Shaoxing wine

Seasoning

1/4 tsp salt
2 tsp crushed rock sugar
2 tsp light soy sauce
2 tsp dark soy sauce

Method

1. Defrost the pork belly in the lower chamber of the refrigerator. Cook in boiling water for about 20 minutes. Remove. Rinse with cold water. Cut into chunks.
2. Heat up a little oil. Stir-fry the ginger and garlic until aromatic. Add the red fermented beancurd and ground bean sauce. Stir-fry until fragrant.
3. Put in the pork belly and stir-fry. Sprinkle with the Shaoxing wine. Finally add the seasoning, 3 cups of water, dried tangerine peel and star aniseed. Simmer for about 1 1/2 hours until the pork belly is tender. Serve.

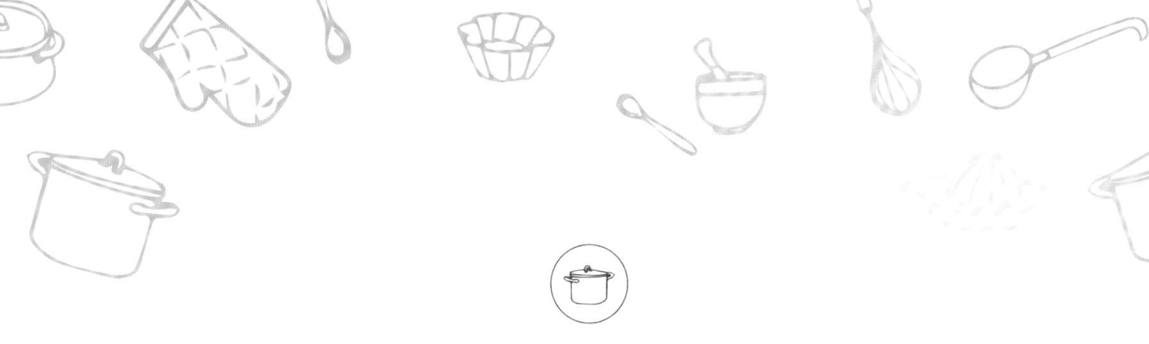

醬燜豬腩肉

Stewed Pork Belly in Miso Tomato Sauce

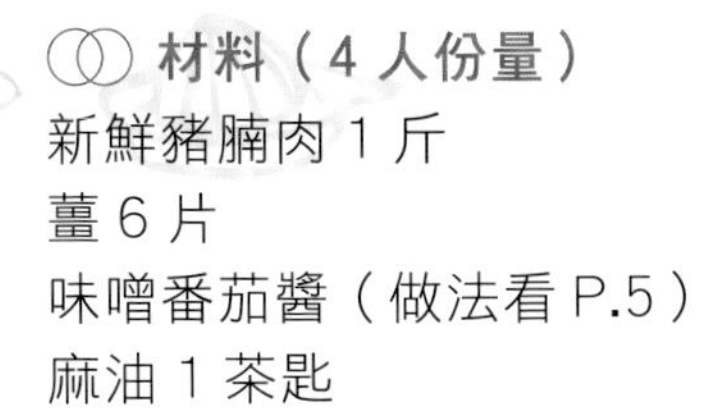

材料（4 人份量）

新鮮豬腩肉 1 斤
薑 6 片
味噌番茄醬（做法看 P.5）
麻油 1 茶匙

Ingredients (Serves 4)

600 g fresh pork belly
6 slices ginger
miso tomato sauce, refer to P.5
1 tsp sesame oil

做法

1. 豬腩肉原塊洗淨，飛水備用。
2. 豬腩肉原塊放入浸過面之滾水內，加入薑片，用中小火煲 45 分鐘，傾入醬汁和麻油 1 茶匙煮滾（注入水蓋過豬腩肉面），用小火煲約 30 分鐘至豬腩肉軟腍。
3. 食用時，可原塊上桌或切成塊食用。

Method

1. Rinse the pork belly and keep it in whole. Blanch it in boiling water and drain.
2. Put the pork belly into a pot. Add boiling water to cover the pork. Put in ginger and bring to the boil. Turn to medium-low heat and simmer for 45 minutes. Add miso tomato sauce and bring to the boil. Add water to cover the pork if needed. Cook over low heat for 30 minutes until the pork is tender.
3. Serve the pork belly in whole or slice it up before serving.

零失敗技巧
Successfully Cooking Skills

豬腩肉為何要原塊燜煮？
保留肉汁，而且以免切塊後的豬腩肉經燜煮而過度縮小。

Why do you cook the pork belly in whole?
It helps keeping the juice in the meat. The meat also shrinks quite a bit in the cooking process. Cutting or slicing the pork at last makes sure the pieces in the right sizes.

如何有效令豬腩肉軟腍？
煲一段時間後，加蓋焗 20 分鐘，然後再煲煮，令豬腩肉容易軟腍，而且節省能源。

How do you make the pork tender?
After cooking for a while, cover the lid and turn off the heat. Leave it for 20 minutes. Then bring to the boil again. Doing this step repeatedly not only saves some fuel, but also makes the pork more tender.

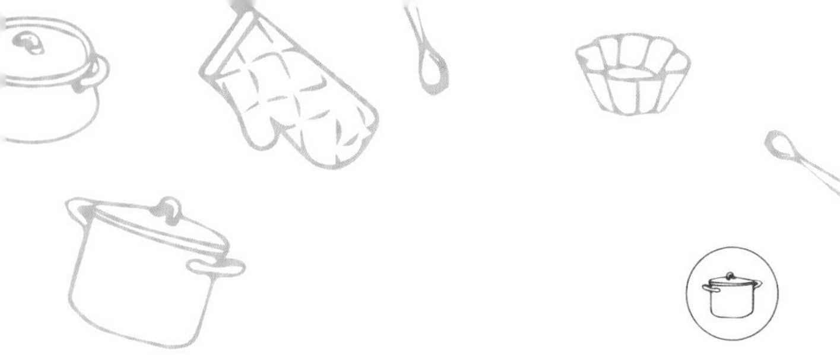

冬菇燒肉燜海參

Stewed Sea Cucumber with Mushroom and Roast Pork

材料（4 人份量）

已浸發海參 6 兩
燒肉 4 兩（切件）
乾冬菇 8 朵
薑 5 片
葱 2 條
紹酒 1 湯匙

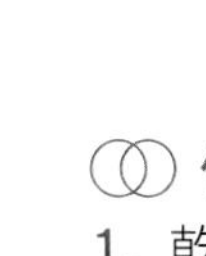

調味料
蠔油 1 湯匙
老抽 1 茶匙
糖半茶匙

芡汁
粟粉 1 茶匙
水 2 湯匙
* 調勻

做法
1. 乾冬菇剪去硬蒂，洗淨，用水浸透，盛起備用（浸冬菇水留用）。
2. 燒滾水，放入海參、薑 2 片及葱 2 條飛水片刻，盛起海參，過冷河，切塊。
3. 燒熱鑊下油 2 湯匙，下薑片及海參炒匀，灒酒，加入燒肉及冬菇拌炒，下調味料及滾水（連浸冬菇水）共 3 杯，轉慢火燜約 45 分鐘，待海參呈軟滑狀，埋芡即成。

零失敗技巧
Successfully Cooking Skills

豬婆參及刺參，食法有何不同？
炮製燜海參菜式，建議選用豬婆參，因肉厚及體型偏大適宜燜煮；刺參肉薄爽滑，宜原隻烹調。

What is the difference in cooking between Poseng and Japanese thorny sea cucumber?
It is recommended to use Poseng in stews as it has a large shape and thick meat. Japanese thorny sea cucumber, which is thin, smooth and crunchy, is suitable for cooking in whole.

長時間燜煮的海參，會糊成一團嗎？
只要浸發時處理妥當，而且海參品質上乘的話，絕對不會糊成一團。

Will the sea cucumber be cooked into a mash through long stewing?
It will not get mashed if it is properly handled during the soaking process and it is of good quality.

浸發海參的步驟繁複，有現成已浸發的海參出售嗎？
若想省掉浸發時間，可向急凍食品店購買已浸發的急凍海參，但緊記挑選有信譽保證的店舖啊！

The soaking process is complicated. Is rehydrated sea cucumber available?
To save time, you may buy frozen rehydrated sea cucumber at frozen food shops, but it is important to select a shop of good reputation!

冬菇燒肉燜海參

Ingredients (Serves 4)

225 g rehydrated sea cucumber
150 g roast pork, cut into pieces
8 dried black mushrooms
5 slices ginger
2 sprigs spring onion
1 tbsp Shaoxing wine

Seasoning

1 tbsp oyster sauce
1 tsp dark soy sauce
1/2 tsp sugar

Thickening glaze

1 tsp cornflour
2 tbsp water
* mixed well

Method

1. Remove the stalks of the dried black mushrooms. Rinse and soak in water thoroughly (reserve the mushroom water). Take out and set aside.
2. Bring water to the boil. Put in the sea cucumber, 2 slices of the ginger and spring onion. Blanch for a while. Remove the sea cucumber. Rinse in cold water. Cut into pieces.
3. Heat up a wok. Add 2 tbsp of oil. Stir-fry the ginger and sea cucumber. Sprinkle with the Shaoxing wine. Put in the roast pork and mushrooms. Stir-fry. Pour in a total of 3 cups of the seasoning and boiling water (with the mushroom water). Turn to low heat and simmer for about 45 minutes. When the sea cucumber is soft, stir in the thickening glaze to finish.

滷水蘿蔔牛腩

Beef Brisket and Radish in Chinese Marinade

材料（4 至 5 人份量）

新鮮牛腩 1 斤
白蘿蔔 1 斤
紹酒 1 湯匙

滷水汁

花椒 1 湯匙
八角 4 粒
陳皮半個
老薑 1 大塊
片糖 3/4 片
鹽 2 茶匙
老抽 4 湯匙
水 6 杯

做法

1. 燒滾滷水汁材料，關火，待 1 小時備用。
2. 牛腩洗淨、飛水，過冷河，瀝乾水分。
3. 白蘿蔔去皮，洗淨，橫切成 2 段，再直切成 4 段，放入滾水內煮 10 分鐘，盛起備用。
4. 燒滾滷水汁，放入牛腩煲滾，轉小火煲半小時，關火，待 1 小時。
5. 再燒滾滷水汁牛腩，轉小火煲半小時，加入白蘿蔔煮滾，關火待 1 小時，盛起牛腩及白蘿蔔，切塊，再放入鍋內，最後傾入紹酒煮滾。

Ingredients (Serves 4-5)

600 g fresh beef brisket
600 g radish
1 tbsp Shaoxing wine

Chinese marinade

1 tbsp Sichuan peppercorns
4 star aniseeds
1/2 dried tangerine peel
1 large piece mature ginger
3/4 slab brown sugar
2 tsp salt
4 tbsp dark soy sauce
6 cups water

Method

1. Bring the Chinese marinade ingredient to the boil. Turn off heat. Leave for 1 hour.
2. Rinse and blanch the beef brisket. Rinse in cold water and drain.
3. Peel and rinse the radish. Cut into half horizontally. Then cut lengthwise into 4 sections. Put into boiling water and cook for 10 minutes. Take out and set aside.
4. Bring the Chinese marinade to the boil. Put in the beef brisket. Bring to the boil. Turn to low heat and cook for 1/2 hour. Turn off heat. Leave for 1 hour.
5. Bring the Chinese marinade with the beef brisket to the boil again. Turn to low heat and cook for 1/2 hour. Add the radish and bring to the boil. Turn off heat. Leave for 1 hour. Take out the beef brisket and radish. Cut into pieces. Return them to the pot. Pour in the Shaoxing wine and bring to the boil. Serve.

零失敗技巧
Successfully Cooking Skills

買不到老薑炮製滷水汁，可以用普通薑塊嗎？

老薑的薑味特別濃郁，增添滷水汁的香氣，建議用大塊老薑。

Mature ginger is not available. Can we use regular ginger to make Chinese marinade?

Mature ginger has a strong flavour which adds fragrance to the Chinese marinade. It is most suitable to use big pieces of mature ginger.

為甚麼牛腩必須要飛水？

飛水後的牛腩可去掉血腥味，燜煮後香濃可口！

Why do you blanch the beef brisket in boiling water first?

Beef brisket needs to be blanched to remove its blood and the gamey taste. The beef brisket will taste better after braised this way.

燜煮牛腩的步驟比較繁複，如何簡化程序？

以焗煮的方法烹調，牛腩軟腍及帶香味；若難以控制複雜的步驟，建議用真空煲烹煮，但緊記上碟前用明火翻滾。

Braising beef brisket is complicated. How to simplify it?

The beef brisket is tender and aromatic when braised. If it is hard to manage, use a pressure cooker but boil it on flame before serving.

咖喱羊腩

Curry Lamb Brisket

材料（4 人份量）

急凍羊腩 600 克
馬鈴薯 2 個（中）
洋葱 1 個
油咖喱 3 湯匙
薑茸、蒜茸及乾葱茸各 1 湯匙
八角 3 粒
肉桂 1 枝
香葉 2 片

調味料

糖半茶匙
鹽適量
胡椒粉少許

做法

1. 羊腩放於雪櫃下層自然解凍，洗淨，抹乾水分。
2. 馬鈴薯刨去外皮，切角；洋葱去外衣，切碎。
3. 燒熱少許油，下馬鈴薯略炸，盛起。
4. 燒熱油 1 湯匙，下羊腩炒透，再加入薑茸、蒜茸、乾葱茸及洋葱炒香，拌入油咖喱拌炒，注入水 2 杯、八角、肉桂及香葉，用小火燜約 1 小時。
5. 最後加入馬鈴薯和調味料再燜約 20 分鐘，待汁液濃稠即可。

Ingredients (Serves 4)

600 g frozen lamb brisket
2 medium potatoes
1 onion
3 tbsp curry paste
1 tbsp finely chopped ginger
1 tbsp finely chopped garlic
1 tbsp finely chopped shallot
3 star aniseeds
1 cinnamon stick
2 bay leaves

Seasoning

1/2 tsp sugar
salt
ground white pepper

Method

1. Defrost the lamb brisket in the lower chamber of the refrigerator. Rinse and wipe it dry.
2. Peel the potatoes. Cut into wedges. Skin the onion and chopped up.
3. Heat up a little oil. Deep-fry the potatoes slightly. Set aside.
4. Heat up 1 tbsp of oil. Stir-fry the lamb brisket thoroughly. Add the ginger, garlic, shallot and onion. Stir-fry until aromatic. Put in the curry paste and stir-fry. Add 2 cups of water, star aniseeds, cinnamon and bay leaves. Simmer over low heat for about 1 hour.
5. Finally add the potatoes and simmer for about 20 minutes until the sauce reduces. Serve.

◎ 零失敗技巧 ◎
Successfully Cooking Skills

羊腩要燜煮多久才入味？

燜煮時間不可少於 1 小時，否則咖喱及香料味難以滲入羊腩內，惹味程度大打折扣！

How long do you braise the lamb brisket for it to be flavourful?

For the best result, braise it for at least one hour. Otherwise, the aromas of the curry and spices won't penetrate into the meat, and the dish won't taste as good.

用急凍羊腩燜煮，容易軟腍嗎？

絕對沒問題！選帶少許油脂的羊腩，口感豐腴，油潤可口，而且產自紐西蘭的羊腩，羊肉味濃重。

Is the frozen lamb brisket easily to be cooked soft?

No problem! Choose lamb brisket with a little fat, which gives a delicious and complicated taste. Also, lamb brisket from New Zealand has intense lamb flavour.

我喜歡吃有口感的馬鈴薯，如何辦？

先將馬鈴薯炸好，燜煮後不易散爛，最後燜煮 20 分鐘即可，別煮太久！

I love chewy potatoes. How to do?

Deep-fry the potatoes before simmering to avoid them falling apart. Do not overcook! Just simmer for 20 minutes.

甚麼是油咖喱？與一般咖喱有何分別？

油咖喱以黃薑、辣椒、洋葱、八角、桂皮及植物油等製成，相比紅咖喱，味道香醇、溫和，而且容易儲存。

What is curry paste? How is it different from regular curry?

Curry paste is made from turmeric, chilli, onion, star aniseed, cinnamon bark, vegetable oil, and more. It has a nicer fragrance and milder flavour compared with red curry. It is also easy for storage.

黃豆苦瓜燜雞

Braised Chicken with Soybean and Bitter Melon

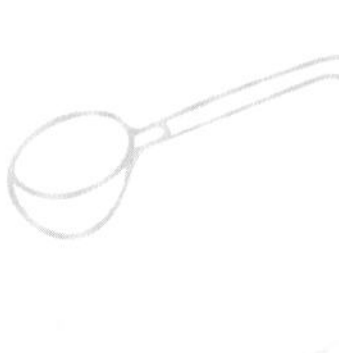

材料（4 人份量）

黃豆 4 湯匙
苦瓜 12 兩
光雞半隻
蒜肉 10 粒
紹酒 1 湯匙

醃料

生抽、薑汁各半湯匙
粟粉 1 茶匙

調味料

生抽半湯匙
鹽、糖各半茶匙
水 2 杯

做法

1. 黃豆用水浸 2 小時，瀝乾水分。
2. 雞洗淨，斬件，加入醃料拌勻，置片刻。
3. 苦瓜開邊，去籽、去瓤，洗淨，切塊。
4. 燒熱鑊，下油 2 湯匙，下蒜肉炒至金黃，加入雞塊炒勻，灒酒，下苦瓜、黃豆及調味料，炒勻後煮滾，轉慢火燜 20 分鐘，至黃豆軟身即成。

Ingredients (Serves 4)

4 tbsp soybeans
450 g bitter melon
1/2 chicken
10 cloves skinned garlic
1 tbsp Shaoxing wine

Marinade

1/2 tbsp light soy sauce
1/2 tbsp ginger juice
1 tsp cornflour

Seasoning

1/2 tbsp light soy sauce
1/2 tsp salt
1/2 tsp sugar
2 cups water

Method

1. Soak the soybeans in water for 2 hours. Drain well.
2. Rinse the chicken and chop into pieces. Mix well with the marinade. Rest for a moment.
3. Cut open the bitter melon. Remove the seeds. Rinse and cut into pieces.
4. Heat a wok. Pour in 2 tbsps of oil. Stir-fry the garlic until golden brown. Add the chicken and stir-fry. Sprinkle with the wine. Add the bitter melon, soybeans and the seasoning. Stir-fry and mix well. Then bring to the boil. Turn to low heat and simmer for 20 minutes until the soybeans turn tender. Serve hot.

零失敗技巧
Successfully Cooking Skills

為甚麼要將黃豆浸泡？

煮黃豆前先用清水浸泡二小時，是讓豆吸收水分至膨脹，能加快將豆煮至軟綿。

Why do you soak the soybeans in water before use?

Soak the soybeans in water for 2 hours before use to let them pick up the moisture and swell. The soybeans will cook and turn tender more quickly this way.

用薑汁醃雞塊有何好處？

可去掉雞肉的血腥味，更能增添香味，一舉兩得。

What is the advantage of marinating chicken with ginger juice?

It is not only removes the bloody smell of the chicken but also improves flavour.

苦瓜味太苦，怎辦？

有以下解決方法：

（1）苦瓜切塊後，用少許鹽醃片刻，壓去苦瓜汁液再沖淨；（2）將苦瓜塊飛水片刻。兩者均可去掉苦瓜的部分苦澀味。

How to reduce the bitter taste of bitter melon?

Two ways: (1) Marinate the sliced bitter melon with a little of salt for a while. Squeeze the juice out and rinse. (2) Scald the sliced bitter melon for a moment.

葱油花雕焗雞球

Braised Chicken with Spices and Hua Diao Wine

材料（4 人份量）

光雞半隻（約 12 兩）
青葱 8 條
乾葱（細）8 粒
薑 2 片
八角 1 粒
紹酒 1 湯匙
清水約 3 湯匙

醃料

鹽 3/4 茶匙
紹興花雕酒 1 1/2 湯匙
糖 1/3 茶匙
老抽半湯匙
蠔油半湯匙
胡椒粉少許

 做法

1. 光雞洗淨、抹乾，斬件，放入醃料拌勻，醃 15 分鐘。
2. 燒熱油 2 湯匙，放入雞球、乾葱、薑片及青葱爆炒，灒紹酒，注入清水及八角，加蓋用中慢火煮 10 分鐘，至雞件熟透及汁液濃稠，即可品嘗。

Ingredients (Serves 4)

1/2 chicken (about 450 g)
8 sprigs spring onion
8 small shallots
2 slices ginger
1 star aniseed
1 tbsp Shaoxing wine
3 tbsp water

Marinade

3/4 tsp salt
1 1/2 tbsp Hua Diao Wine
1/3 tsp sugar
1/2 tbsp dark soy sauce
1/2 tbsp oyster sauce
ground white pepper

Method

1. Rinse the chicken. Wipe it dry. Chop into pieces. Mix with the marinade and rest for 15 minutes.
2. Heat up 2 tbsp of oil. Stir-fry the chicken, shallots, ginger and spring onion. Sprinkle with the Shaoxing wine. Add the water and star aniseed. Cook over low-medium heat for 10 minutes with a lid on. When the chicken is fully cooked and the sauce thickens, dish up and serve.

零失敗技巧
Successfully Cooking Skills

如何令葱油香氣徹底散發？

先將青葱及乾葱等料頭徹底爆香，才放入雞球拌炒，葱油香氣就能充份滲入雞肉內！

How to make the spices spread their fragrance?

Stir-fry the spices like spring onion and shallots first. When you smell the fragrance, add the chicken. The aromatic flavour will completely infuse into the chicken.

可以用其他料酒代替花雕酒嗎？

用花雕酒焗煮的雞球，酒香四溢，別的料酒難以代替。

Can we use a wine substitute?

Hua Diao wine will give the chicken a strong aromatic wine flavour, and can hardly be replaced.

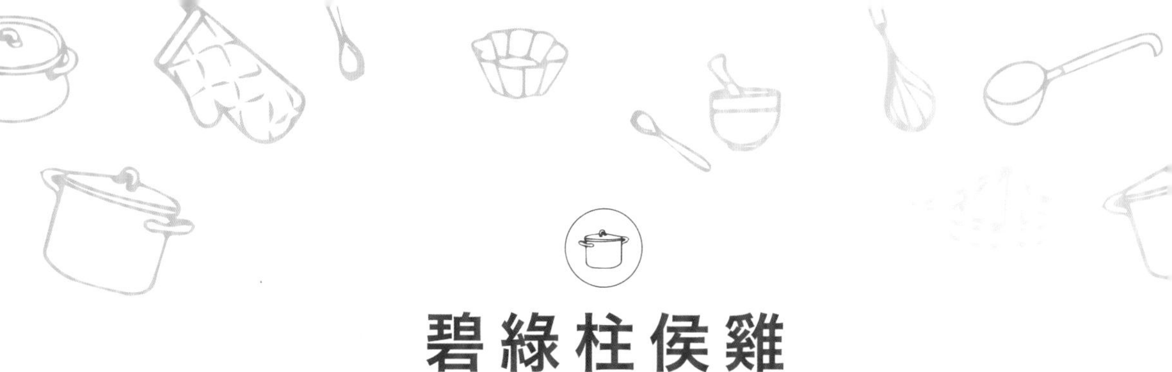

碧綠柱侯雞

Braised Chicken in Chu Hou Sauce

材料（4人份量）

光雞半隻（約12兩）
柱侯醬 1 1/2 湯匙
小棠菜 5 兩
蒜頭 1 粒（切片）
薑 2 片
紹酒 1 茶匙

醃料

生抽 2 茶匙
薑汁酒 1 湯匙
生粉半湯匙
油 1 湯匙

調味料
水 3 杯
鹽 1 茶匙
糖半茶匙

汁料
水 125 毫升
生抽半湯匙
糖半茶匙
麻油少許

生粉芡
生粉 1 茶匙
水 1 湯匙

做法
1. 光雞洗淨，斬件，抹乾水分，加入醃料拌勻醃 15 分鐘。
2. 小棠菜修剪妥當，放入已下油的滾水內，加入調味料灼熟，盛於碟上。
3. 熱鑊下油 1 湯匙，下雞件煎炒至金黃色，盛起。
4. 下少許油，爆香蒜片及柱侯醬，放入薑片及雞件炒勻，灒紹酒，傾入汁料煮滾，轉慢火煮約 10 分鐘至雞件熟透，埋芡，盛於小棠菜上面。

零失敗技巧
Successfully Cooking Skills

爆炒蒜片及柱侯醬，有何注意之處？
建議調至細火爆炒，否則醬料及蒜片容易焦燶。
What should be noted when stir-frying garlic and Chu Hou sauce?
Use low heat to stir-fry to avoid the sauce and garlic getting burnt.

可用其他醬料代替柱侯醬嗎？
嗜辣者可用XO醬；或用磨豉醬、蒜茸豆豉醬炒煮，另有一番風味！
Can we use a sauce substitute?
Use XO sauce if you are fond of spicy food. Ground bean sauce or black bean garlic sauce is also wonderful!

炒煮雞件有何技巧？
雞件別太大件，否則難以熟透。此外，加入汁料後必須調慢火焗煮，令雞件入味及全熟。
What is the cooking tips on handling the chicken?
Cut the chicken into small pieces to make it cook through easily. After adding the sauce, cook over low heat with a lid on to let the flavour penetrate the chicken and make it fully done.

Ingredients (Serves 4)

1/2 chicken (about 450 g)
1 1/2 tbsp Chu Hou sauce
188 g Shanghai white cabbage
1 clove garlic, sliced
2 slices ginger
1 tsp Shaoxing wine

Marinade

2 tsp light soy sauce
1 tbsp ginger juice wine
1/2 tbsp caltrop starch
1 tbsp oil

Seasoning

3 cups water
1 tsp salt
1/2 tsp sugar

Sauce

125 ml water
1/2 tbsp light soy sauce
1/2 tsp sugar
sesame oil

Caltrop starch solution

1 tsp caltrop starch
1 tbsp water

Method

1. Rinse the chicken. Chop into pieces. Wipe them dry. Mix with the marinade and rest for 15 minutes.
2. Trim the Shanghai white cabbage. Put in boiling water added with oil. Pour in the seasoning. Blanch until cooked. Arrange on a plate.
3. Add 1 tbsp of oil in a heated wok. Put in the chicken and fry until golden. Set aside.
4. Put in a little oil. Stir-fry the garlic and Chu Hou sauce until aromatic. Add in the ginger and chicken. Stir-fry and toss well. Sprinkle with Shaoxing wine. Pour in the sauce and bring to the boil. Turn to low heat and cook for about 10 minutes until the chicken is cooked through. Mix in the caltrop starch solution. Arrange on the Shanghai white cabbage.

豉汁涼瓜燜雞

Stewed Chicken and Bitter Melon in Fermented Black Bean Sauce

材料（4 人份量）

光雞 8 兩
涼瓜 6 兩
豆豉 1 湯匙
蒜茸 2 茶匙
紅椒 1 隻（切圈）
紹酒 1 湯匙

醃料

生抽 2 茶匙
老抽半茶匙
糖 1/8 茶匙
薑汁半湯匙
生粉半湯匙
油半湯匙

調味料

水 5 湯匙
生抽 1 茶匙
蠔油半湯匙
糖半茶匙
鹽 1/3 茶匙

做法

1. 光雞洗淨，抹乾水分，斬件，加入醃料拌勻醃 15 分鐘。
2. 涼瓜及紅椒去瓤、切片。
3. 熱鑊下油，加入涼瓜爆炒，下鹽半茶匙、水 2 湯匙拌炒至水分收乾，盛起。
4. 下油用慢火爆香豆豉及蒜茸，下雞件炒勻，灒紹酒，加蓋焗煮 3 分鐘，加入調味料及涼瓜續煮 5 分鐘，最後下紅椒圈拌勻上碟。

零失敗技巧
Successfully Cooking Skills

刮苦瓜瓤有何注意之處？
除去掉苦瓜籽外，緊記要刮深一點，連淺綠色的部份也一併刮去，以免影響口感！
What should be noted when scraping the pulp from bitter melon?
Scrape deeper into the pulp, removing the seeds and also the part in light green to avoid unpleasant taste!

想吃爽脆的苦瓜，如何辦？
苦瓜切片後，用少許鹽拌勻醃 5 分鐘，沖去鹽分，瀝乾水後快炒，苦瓜會更爽脆！
I love crunchy bitter melons. How to do?
To make them crunchy, marinate the sliced bitter melons with a pinch of salt for 5 minutes. Then rinse, drain, and stir-fry quickly.

Ingredients (Serves 4)

300 g chicken
225 g bitter melon
1 tbsp fermented black beans
2 tsp finely chopped garlic
1 red chilli, cut into rings
1 tbsp Shaoxing wine

Marinade

2 tsp light soy sauce
1/2 tsp dark soy sauce
1/8 tsp sugar
1/2 tbsp ginger juice
1/2 tbsp caltrop starch
1/2 tbsp oil

Seasoning

5 tbsp water
1 tsp light soy sauce
1/2 tbsp oyster sauce
1/2 tsp sugar
1/3 tsp salt

Method

1. Rinse the chicken. Wipe it dry. Chop into pieces. Mix with the marinade and rest for 15 minutes.
2. Remove the pith of the bitter melon and red chilli. Slice.
3. Add oil in a heated wok. Stir-fry the bitter melon. Add 1/2 tsp of salt and 2 tbsp of water. Stir-fry until dry. Dish up.
4. Put oil in the wok. Stir-fry the fermented black beans and garlic over low heat until fragrant. Add the chicken and stir-fry. Sprinkle with the Shaoxing wine. Cook with a lid on for 3 minutes. Put in the seasoning and bitter melon. Cook for another 5 minutes. Finally add the red chilli and mix well. Serve.

海南雞伴油飯

Hainanese Chicken with Flavoured Oil Rice

材料（4 至 6 人份量）
冰鮮雞 1 隻（大）
白米 2 杯
黑醬油適量

醃料
南薑及薑各數片
乾葱頭 2 粒
鹽 2 茶匙

油飯配料
雞油膏 1 塊
斑蘭葉 2 塊
香茅 2 枝
薑 6 片
蒜肉 10 粒
雞湯 1 杯
水 1 杯
鹽少許

薑蒜辣汁
紅辣椒幼粒 2 茶匙
薑茸 1 1/2 茶匙
蒜茸 2 湯匙
鹽半茶匙
滾油 2 湯匙
酸柑汁（泰國青檸汁）2 湯匙
糖半茶匙

薑蒜辣汁做法
紅辣椒粒、薑茸及蒜茸拌勻，灒滾油，加入餘下材料拌勻，備用。

海南雞做法
1. 雞洗淨，去內臟，用鹽於雞內外擦勻。
2. 將醃料拍碎，醃雞約 2 小時，再將醃料放入雞腔內。
3. 燒滾水（宜浸過雞面），放入雞及鹽 1 湯匙，滾起關火，浸 15 分鐘。
4. 再開火滾起，關火，再浸 15 分鐘。
5. 雞取出，浸於冰水待片刻，斬件。

油飯做法
1. 白米用水浸約 1 小時，瀝乾水分。
2. 雞油膏略煎，待雞油溢出後，下蒜肉略炒，加入白米炒勻，轉放飯煲內。
3. 加入其餘的配料，拌勻，按掣煲煮成油飯，煮熟後取出配料，伴海南雞、薑蒜辣汁及黑醬油享用。

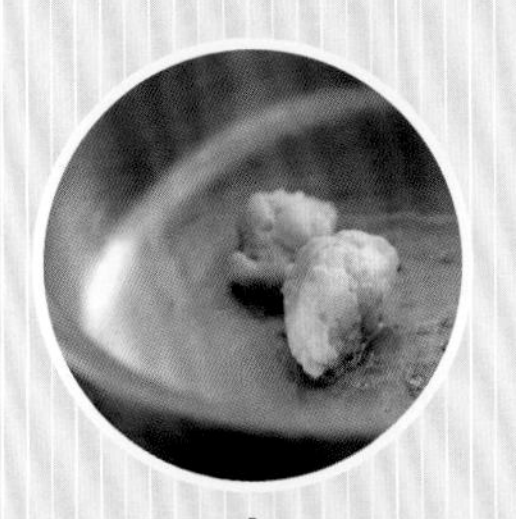
1

2

3

4

Ingredients (Serves 4-6)

1 large chilled chicken
2 cups rice
thick soy sauce

Marinade

a few slices galangal
a few slices ginger
2 shallots
2 tsp salt

Condiments for oil rice

1 piece chicken fat
2 pandan leaves
2 stalks lemongrass
6 slices ginger
10 cloves skinned garlic
1 cup chicken stock
1 cup water
salt

Ginger garlic chilli sauce

2 tsp finely diced red chilli
1 1/2 tsp finely chopped ginger
2 tbsp finely chopped garlic
1/2 tsp salt
2 tbsp hot oil
2 tbsp calamansi juice (Thai lime juice)
1/2 tsp sugar

Method for ginger garlic chilli sauce

Mix the red chilli, ginger and garlic well. Pour in the hot oil. Add the rest ingredients and mix well. Set aside.

Method for Hainanese chicken

1. Rinse and gut the chicken. Rub the inside and outside of the chicken with salt.
2. Crush the marinade. Marinate the chicken for about 2 hours. Then put the marinade into the chicken cavity.
3. Bring water to the boil (water enough to cover the chicken). Put in the chicken and 1 tbsp of salt. Bring to the boil. Turn off heat. Soak for 15 minutes.
4. Turn on heat and bring to the boil. Turn off heat. Soak for 15 minutes.
5. Remove the chicken. Soak in iced water for a moment. Chop into pieces.

Method for flavoured oil rice

1. Soak the rice in water for about 1 hour. Drain.
2. Slightly fry the chicken fat until it releases oil. Put in the garlic and stir-fry slightly. Add the rice and stir-fry evenly. Transfer to a rice cooker.
3. Put in the rest condiments. Mix well. Press the button and cook into oil rice. When it is done, remove the condiments. Serve with the chicken, ginger garlic chilli sauce and thick soy sauce.

零失敗技巧
Successfully Cooking Skills

為甚麼浸雞時要下少許鹽？

浸雞時加入少許鹽，能夠緊鎖肉質，保存濃郁的雞味。

Why do you add salt to the water before soaking the chicken in it?

Adding a pinch of salt to the water helps seal in the juices and keeps the flesh succulent. It also retains the meaty flavour in the flesh.

為何重複開火及關火的步驟？

用這個方法浸煮的雞，縱使用上冰鮮雞烹調，也可吃出嫩滑的質感。

Why repeat turning on and off the heat?

By doing so, the cooked chicken will taste smooth even though it is a chilled chicken.

必須用這麼多配料炒油飯嗎？

絕對是！油飯會滲有斑蘭葉及香茅的清香之味。

Is it necessary to cook the oil rice with so many condiments?

Yes, absolutely! The rice will smell the light fragrance of pandan leaves and lemongrass.

甚麼是雞油膏？有何作用？

雞油膏是雞的油脂，可在雞皮下切出，用以爆香油飯的配料，雞味香濃，令你多吃幾碗！

What is chicken fat? What is its use?

It is the fat of chicken underneath the skin from which it can be cut out. The fat is used as a condiment to be stir-fried with the rice to give it a strong chicken flavour, which is mouthwatering!

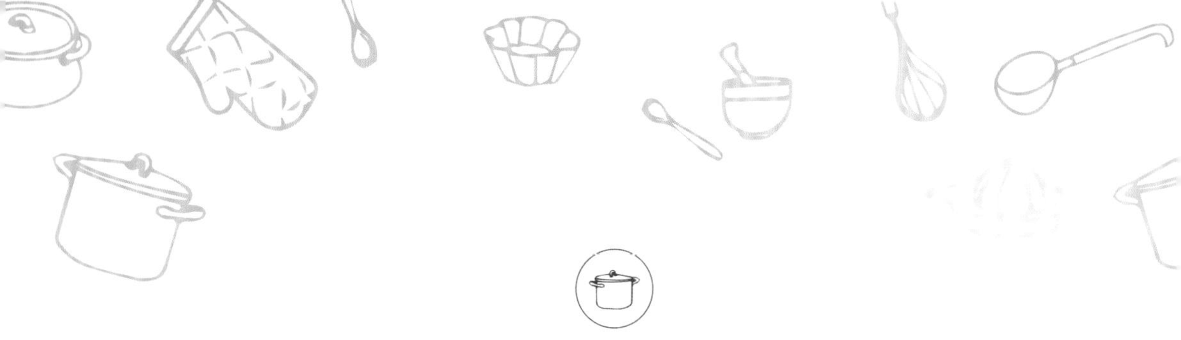

蝦乾鮫魚餅

Fried Mackerel and Dried Shrimp Patties

材料（4 人份量）

鮫魚肉 8 兩
蝦乾半兩
韭菜粒 3 湯匙

調味料

鹽半茶匙
魚露 2 茶匙
胡椒粉少許
浸泡蝦乾水 2 湯匙

做法

1. 鮫魚肉用匙羹依逆紋刮出魚肉。
2. 蝦乾洗淨，浸軟，切粒（浸泡蝦乾水留用）。
3. 鮫魚茸、蝦乾、韭菜粒及調味料拌勻，順一方向攪拌至起膠，分成 15 等份，搓成魚餅。
4. 燒熱油，下魚餅用慢火煎至金黃色及熟透，上碟即成。

零失敗技巧
Successfully Cooking Skills

除用扁鮫魚肉外，還可用甚麼魚？

可選用鯪魚或其他品種的鮫魚，但扁鮫魚的肉質及膠質皆最適宜，令魚餅更具彈性。

What other fish can be used except the flat mackerel?

Dace or other kind of mackerel can be used. But the texture of flat mackerel is most suitable for fried fish patties as it is elastic and chewy.

一尾鮫魚可刮出多少魚茸？

一尾約 1.5 斤重之鮫魚，可刮出約 1 斤重魚茸。購買時可請魚販代起魚肉。

How much flesh can be obtained from a mackerel?

A mackerel weighed 900 g can give about 600 g of flesh. You can simply ask the fish monger for boning.

加入浸泡蝦乾水有何作用？

泡浸後的清水帶蝦乾之鮮香味，別浪費，當調味料加入魚餅內，加倍提升魚餅的鮮味。

What's the point of adding water from soaking dried shrimps?

Don't waste the water from soaking dried shrimps as it has fresh shrimp taste. It can enhance the flavour of the fish patties as the seasoning.

蝦乾鮫魚餅

Ingredients (Serves 4)

300 g boned mackerel
19 g dried shrimps
3 tbsp diced Chinese chive

Seasoning

1/2 tsp salt
2 tsp fish sauce
ground white pepper
2 tbsp water from soaking dried shrimps

Method

1. Scoop flesh from mackerel with a spoon along the vein in a reverse direction.
2. Rinse dried shrimps. Soak until soft and dice. (Set aside the water from soaking dried shrimps.)
3. Mix the mackerel flesh, dried shrimps, diced Chinese chive and seasoning. Stir in a direction until sticky. Divide into 15 portions and knead into fish patties.
4. Heat oil in wok. Fry the fish patties over low heat until golden brown and done. Serve.

惹味蝦

Braised Prawns with Sweet, Sour and Spicy Sauce

材料（3 人份量）

中蝦 8 兩
薑米 1 湯匙
葱 2 條
紹酒 1 湯匙

汁料

水及鎮江醋各 1 1/2 湯匙
糖 1 湯匙
辣椒油半湯匙
鹽 1/4 茶匙
麻油 1 茶匙

做法

1. 修剪中蝦的蝦鬚及蝦腳，挑去蝦腸，洗淨，抹乾水分。
2. 葱分成葱白及葱段。
3. 熱鑊下油，加入薑米及葱白爆香，放入中蝦煎至剛轉成紅色，灒紹酒，傾入汁料，加蓋焗片刻。
4. 焗煮至汁料濃稠，最後灑入葱段拌炒，上碟。

零失敗技巧
Successfully Cooking Skills

蝦肉會煮得過熟嗎？

中蝦煎至半熟（剛轉成紅色），注入汁料同煮，蝦肉不會煮得過熟。

Would prawns be over-cooked?

Fry medium prawns until medium cooked (just turn red). Pour in the sauce and cook it together with the prawns hence the prawns would not be over-cooked.

若想方便進食，烹調前可先剝掉蝦殼嗎？

絕對可以，但汁料份量需略調校，因蝦隻已去掉外殼，味道會直接滲入蝦肉。

Can the prawns be shelled before cooking for easy consumption?

Absolutely but the sauce need to be adjusted. Since the prawns have been shelled, any taste would go into the prawns directly.

Ingredients (Serves 3)

300 g medium-sized prawns
1 tbsp chopped ginger
2 sprigs spring onion
1 tbsp Shaoxing wine

Seasoning

1 1/2 tbsp water
1 1/2 tbsp Zhenjiang vinegar
1 tbsp sugar
1/2 tbsp chilli oil
1/4 tsp salt
1 tsp sesame oil

Method

1. Trim prawn tentacles and legs. Devein, rinse and wipe dry.
2. Cut spring onion into white parts and cut the green parts into sections.
3. Add oil into a hot wok. Put in chopped ginger and white part of spring onion. Stir-fry quickly until fragrant. Add medium prawns and fry until they just turn red. Sizzle in Shaoxing wine. Pour in sauce and cover the lid.
4. Cook for a while until the sauce thickens. Sprinkle over sectioned spring onion and stir-fry well. Serve.

梅子蜜汁烤雞髀

Roasted Chicken Legs in Plum and Honey Sauce

材料（4 人份量）

雞髀 4 隻（約 800 克）
酸梅 2 粒
韓國蜂蜜梅子茶適量
（塗面用）

醃料

鹽半茶匙
生抽 1 1/2 湯匙
韓國蜂蜜梅子茶 3 湯匙

做法

1. 酸梅去核，剁蓉備用。
2. 雞髀洗淨，吸乾水分，用竹籤在外皮戳上孔，加入醃料及酸梅茸拌勻醃 2 小時。
3. 預熱焗爐 200℃；錫紙鋪焗盤上，塗上油。
4. 雞髀放於焗盤內，放入焗爐烤約 20 分鐘，取出，塗上蜂蜜梅子茶，轉至 180℃烤 6 分鐘，上碟趁熱享用。

Ingredients (Serves 4)

4 chicken legs (about 800 g)
2 pickled plums
Korean honey plum tea (for coating)

Marinade

1/2 tsp salt
1 1/2 tbsp light soy sauce
3 tbsp Korean honey plum tea

Method

1. Remove the seed of pickled plums. Finely chopped. Set aside.
2. Rinse the chicken legs. Wipe dry. Pierce holes in the skin with a bamboo skewer. Mix with the marinade and the pickled plum. Rest for 2 hours.
3. Preheat an oven to 200°C. Line an aluminium foil on a baking tray. Brush with oil.
4. Put the chicken legs on the baking tray. Place in the oven and bake for about 20 minutes. Remove. Spread the honey plum tea on the chicken legs. Adjust to 180°C and bake for 6 minutes. Serve hot.

零失敗技巧
Successfully Cooking Skills

在整隻雞髀的外皮戳上小孔，有何作用？

戳上小孔後，醃料容易滲入雞肉，肉質帶陣陣梅子香氣；或可於雞髀最厚肉部分戳上孔也可。

Is it necessary to pierce small holes into the skin of the whole chicken leg?

The marinade will easily infuse into the meat through the small holes, giving it the nice fragrance of plum. It will also work by piercing holes just in the thickest part of the meat.

為甚麼還需要配酸梅烹調？

配上中菜用的酸梅，剁成茸，更能提升梅子的味道，更有層次！

Why cook with pickled plums?

To cook with finely chopped pickled plums used in Chinese cuisine will enhance the plum taste of the dish, bringing the savour to a higher level!

黑毛豬腩金菇卷

Enokitake Mushrooms Rolled in Kurobuta Pork Belly

材料

黑毛豬腩片 8 片
金菇 1 包
甘筍 2 兩
葱 4 條
味噌番茄醬（做法看 P.5）

做法

1. 金菇切去尾端，洗淨，隔去水分。
2. 甘筍去皮，洗淨，切絲；葱去鬚根，洗淨，切段。
3. 醬汁以小火煮滾，備用。
4. 取一片黑毛豬腩片，鋪平，排上適量金菇、甘筍及葱段，緊緊捲成金菇卷（最後抹上少許乾粟粉黏緊）。
5. 金菇卷放入煎鑊內，煎至全熟及金黃色，上碟，澆上味噌番茄醬伴吃。

Ingredients

8 slices Kurobuta pork belly
1 pack enokitake mushrooms
75 g carrot
4 sprigs spring onion
miso tomato sauce, refer to P.5

Method

1. Cut off the roots of the enokitake mushrooms. Rinse well. Drain.
2. Peel and rinse the carrot. Shred it. Set aside. Cut off the fibrous roots of the spring onion. Rinse well and cut into short lengths.
3. Cook the miso tomato sauce over low heat until it boils. Set aside.
4. Lay flat a slice of pork belly. Arrange some enokitake mushrooms, carrot and spring onion on it. Roll the pork up tightly. You may secure the seam by dusting some cornflour on the pork.
5. Put the pork belly rolls into a pan with a little oil. Fry until the rolls are done and the pork is golden. Arrange on a plate. Drizzle with the miso tomato sauce from step 3. Serve.

零失敗技巧
Successfully Cooking Skills

煎金菇卷時，餡料容易熟透嗎？
灑入少許水分，加蓋焗煮，容易熟透，再煎至水分收乾即可。

Does the filling get properly cooked when you fry the rolls in a pan like that?

Sprinkle some water halfway through and cover the lid. That would ensure the filling is properly cooked. Just keep on frying until the liquid dries out.

酸梅椒醬蒸魚雲

Steamed Fish Head with Sour Plums and Chillies in Soybean Sauce

材料（3至4人份量）

大魚魚頭12兩
酸梅3粒
指天椒3隻
磨豉醬2茶匙
蒜茸2茶匙
薑絲2湯匙
葱粒2湯匙

調味料

鹽1/4茶匙
糖1湯匙
老抽1茶匙

做法

1. 大魚頭切成6至8件（魚販可代勞），洗淨，瀝乾水分。
2. 酸梅去核，剁成茸。指天椒去籽，切絲。
3. 調味料、酸梅茸、指天椒、磨豉醬、蒜茸及薑絲拌勻。
4. 將上述之調味料與大魚頭拌勻，放入碟內，隔水蒸8分鐘，灑入葱粒，再蒸半分鐘即成。

零失敗技巧
Successfully Cooking Skills

如何選購大魚頭？
必須選購新鮮的大魚頭，否則帶魚腥味。

How to choose big fish head?
It is a must to choose fresh big fish head or else it brings fishy smell.

通常買一瓶磨豉醬回來，很久也吃不完，怎辦？
可到雜貨店購買散裝的磨豉醬，可隨每次用量購買，毋須囤積於雪櫃。

How to handle the remaining bottle of ground soybean sauce?
You can just buy the amount of ground soybean sauce you need each time at the grocery store so that a whole bottle need not be stored in the refrigerator.

Ingredients (Serves 3-4)

450 g head of bighead carp
3 sour plums
3 bird's eye chillies
2 tsp ground soybean sauce
2 tsp minced garlic
2 tbsp shredded ginger
2 tbsp diced spring onion

Seasoning

1/4 tsp salt
1 tbsp sugar
1 tsp dark soy sauce

Method

1. Cut fish head into 6 to 8 pieces (or ask the fish monger for help). Rinse and drain.
2. Stone sour plums and chop into minced form. Seed bird's eye chillies and shred.
3. Mix the seasoning, minced sour plums, bird's eye chillies, ground soybean sauce, minced garlic and ginger shreds.
4. Mix the above seasoning with the fish head. Put into a plate and steam for 8 minutes. Sprinkle over spring onion dices and steam for 1/2 minute more. Serve.

蒸腐皮魚蝦卷

Steamed Beancurd Skin Roll with Fish and Shrimp Filling

材料（4 人份量）

鮮腐皮 1 塊
鯪魚膠 3 兩
鮮蝦仁 3 兩
菜遠適量

醃料

胡椒粉少許
鹽 1/4 茶匙
蛋白半個
麻油、粟粉各 1 茶匙

調味料

生抽半湯匙
熟油 1 湯匙

做法

1. 蝦仁洗淨，瀝乾水分，用刀拍扁，剁成粗粒，下醃料拌勻，加入鯪魚膠順一方向拌至起膠。
2. 菜遠洗淨，灼熟，瀝乾水分，備用。
3. 鮮腐皮剪去硬邊，用乾淨濕毛巾略抹，鋪上魚蝦膠，捲成長條形，隔水用大火蒸 10 分鐘，切塊，伴上菜遠，灒調味料即成。

零失敗技巧

Successfully Cooking Skills

為甚麼要將蝦仁剁成粗粒？
切勿將蝦仁剁得太幼細，會失去咬感。
Should we coarsely chop shelled shrimps?
Do not finely chop the shrimps to keep them chewy.

於腐皮抹上厚厚的魚蝦膠嗎？
建議抹上厚厚的魚蝦膠，咬入口，魚肉及蝦仁的質感豐腴美味。
Is it necessary to spread a thick layer of fish and shrimp paste on beancurd skin?
A thick layer has a rich texture of fish and shrimp to make the dish yummy.

鮮腐皮毋須用水清洗嗎？
毋須清洗，只需用濕布略抹即可。
We need not rinse the fresh beancurd skin. Is that right?
Yes. It can be cleaned by slightly wiping with a damp towel.

蒸腐皮魚蝦卷

Ingredients (Serves 4)

1 piece fresh beancurd skin
113 g minced dace
113 g shelled fresh shrimps
flowering Chinese cabbages

Marinade

ground white pepper
1/4 tsp salt
1/2 egg white
1 tsp sesame oil
1 tsp cornflour

Seasoning

1/2 tbsp light soy sauce
1 tbsp cooked oil

Method

1. Rinse and drain the shelled shrimps. Pound with a knife and coarsely dice. Mix well with the marinade. Add the minced dace and stir in one direction until sticky.
2. Rinse the flowering Chinese cabbages. Blanch until done. Drain and set aside.
3. Cut away the hard edge of the beancurd skin. Slightly wipe it with a clean damp towel. Spread the fish and shrimp paste on the skin. Roll into a strip. Steam over high heat for 10 minutes. Cut into pieces. Drizzle the seasoning on top. Serve with the flowering Chinese cabbages.

家鄉蒸蠔缽

Steamed Oyster with Egg and Pork in Earthen Bowl

材料（4 人份量）

桶蠔 450 克
免治豬肉 80 克
冬菇 4 朵（浸軟、去蒂，切絲）
芹菜碎 2 湯匙
葱花 1 湯匙
雞蛋 5 個
上湯半杯
陳皮 1 角（浸軟、刮瓤，切幼絲）
炸蒜片 1 1/2 湯匙

醃料（蠔）

鹽 1/4 茶匙
紹酒半茶匙
胡椒粉少許

醃料（免治豬肉）

鹽、糖、生抽、紹酒
及粟粉各半茶匙
胡椒粉少許

調味料

鹽半茶匙
胡椒粉少許

做法

1. 蠔用少許鹽及粟粉擦淨，沖洗乾淨，飛水，吸乾水分，切件，與醃料拌勻。
2. 免治豬肉與醃料拌勻，備用。
3. 燒熱少許油，下豬肉略炒，加入冬菇絲炒勻，盛起。
4. 雞蛋拂勻，注入上湯及調味料拌勻，加入其他材料（蠔除外）。
5. 瓦缽內掃上少許油，放入蠔件，拌入雞蛋漿，隔水中火蒸 12 分鐘。
6. 將瓦缽放入預熱焗爐，烤至表面呈微黃色即可。

Ingredients (Serves 4)

450 g shucked oysters
80 g minced pork
4 dried black mushrooms, soaked to soften; stalks removed; shredded
2 tbsp chopped Chinese celery
1 tbsp diced spring onion
5 eggs
1/2 cup stock
1 slice dried tangerine peel, soaked to soften; pith scraped off; finely shredded
1 1/2 tbsp deep-fried garlic slices

Marinade (oysters)

1/4 tsp salt
1/2 tsp Shaoxing wine
ground white pepper

Marinade (minced pork)

1/2 tsp salt
1/2 tsp sugar
1/2 tsp light soy sauce
1/2 tsp Shaoxing wine
1/2 tsp cornflour
ground white pepper

Seasoning

1/2 tsp salt
ground white pepper

Method

1. Rub the oysters clean with a little salt and cornflour. Rinse and scald. Wipe dry and cut into pieces. Mix with the marinade.
2. Mix the minced pork with the marinade. Set aside.
3. Heat up a little oil. Stir-fry the minced pork slightly. Add the black mushrooms and stir-fry evenly. Set aside.
4. Beat the eggs. Pour in the stock and seasoning. Mix well. Add the other ingredients (except the oysters).
5. Brush a little oil on an earthen bowl. Put in the oysters. Pour in the egg mixture from step 4. Steam over medium heat for 12 minutes.
6. Put the earthen bowl in a preheated oven. Bake until the surface is light brown to finish.

零失敗技巧
Successfully Cooking Skills

桶蠔容易擦洗嗎？

桶蠔比鮮蠔容易擦洗，最重要將桶蠔的群邊徹底洗淨，以免藏有微細的砂粒。。

Is it easy to clean shucked oysters packed in bottles?

They are easily to be cleaned compared with fresh oysters. It is most important to clean their ruffle-like edges thoroughly to remove hidden sand grains.

害怕烤焗食物容易上火，刪掉烤焗步驟可以嗎？

蠔缽蒸熟後可直接享用，放入焗爐只是將表面烤得香脆一點，多一份口感！

According to Chinese medicine, taking baked food easily induces heat in the body. Can baking be skipped?

This dish can be served immediately after steamed. Baking can make the surface crunchy, giving an additional taste!

陳皮為何要刮去內瓤？

陳皮內瓤屬寒涼，而且令餸菜帶淡淡的苦味，必須刮去。

Why scrape the pith off dried tangerine peel?

According to Chinese medicine, the pith of dried tangerine peel is cooling food. It also gives the dish a light bitter taste and so it must be removed.

金銀蒜蒸象拔蚌仔

Steamed Geoduck Clams with Garlic

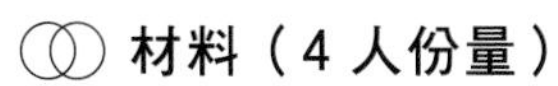

材料（4 人份量）
象拔蚌仔 8 隻
蒜茸 3 湯匙
乾粉絲 1 兩
指天椒 3 隻
葱粒 1 湯匙

調味料
清雞湯 125 毫升
鹽半茶匙
頭抽 1 湯匙

做法
1. 象拔蚌仔去掉腸臟（魚販可代勞），洗淨，瀝乾水分。
2. 取 1/3 份量蒜茸炸香成金蒜，備用。
3. 燒熱清雞湯，下粉絲浸軟，剪成段。
4. 指天椒切圈；用油及將餘下之蒜茸炒香，下鹽拌勻，待涼成香蒜。
5. 象拔蚌仔排於碟上，放上適量粉絲、香蒜及辣椒粒，隔水用中慢火蒸 4 分鐘，灑入葱粒，再蒸一會，取出。
6. 煮滾油 1 湯匙，澆在象拔蚌上，下頭抽，灑入金蒜，趁熱享用。

零失敗技巧
Successfully Cooking Skills

家裏沒有頭抽，怎辦？
可用生抽 1 湯匙及糖半茶匙煮溶代替。
What can I do if I do not have premium soy sauce?
You can cook 1 tbsp of light soy sauce and 1/2 tsp of sugar until molten to replace it.

象拔蚌仔蒸 4 分鐘熟透嗎？
絕對熟透，烹調過久會令肉質變韌。
Can geoduck clams be done by steaming 4 minutes only?
Absolutely. Increase the cooking time only makes them harder in texture and not crunchy.

最後必須澆上熟油嗎？
澆上熟油，令頭抽不會直接滲入象拔蚌肉，以免鹹味太重。
Is it a must to pour over cooked oil at the last?
Before you add premium soy sauce, pour over cooking oil first in order to avoid too heavy salty taste.

金銀蒜蒸象拔蚌仔

Ingredients (Serves 4)

8 small geoduck clams
3 tbsp minced garlic
38 g dried vermicelli
3 bird's eye chillies
1 tbsp diced spring onion

Seasoning

125 ml chicken broth
1/2 tsp salt
1 tbsp premium soy sauce

Method

1. Remove entrails from small geoduck clams (or ask the fish monger for help). Rinse and drain.
2. Deep-fry 1/3 portion of minced garlic until fragrant and set aside.
3. Heat chicken broth and add in vermicelli to soak until soft. Cut into sections.
4. Cut bird's eye chillies into rings. Stir-fry the remaining minced garlic with oil until fragrant. Put in salt and mix well. Set aside to let cool.
5. Arrange small geoduck clams into a plate. Put in vermicelli, stir-fried minced garlic and diced chillies. Steam over medium-low heat for 4 minutes. Sprinkle over diced spring onion and steam for a while.
6. Bring 1 tbsp of oil to the boil. Pour over the small geoduck clams and add premium soy sauce. Sprinkle over deep-fried minced garlic and serve hot.

黑糯米酒紅棗蒸雞

Steamed Chicken with Red Dates in Black Glutinous Rice Wine

材料（4 人份量）

上等冰鮮雞 1 隻
紅棗 15 粒
冰糖 2 湯匙（舂碎）
黑糯米酒 1 杯
鹽 1 茶匙
薑 6 片

做法

1. 紅棗去核、洗淨，備用。雞洗淨，瀝乾水分。
2. 雞內腔用鹽抹勻，將半份紅棗、半份冰糖、薑片及少許黑糯米酒放入雞腔內，餘下半份之紅棗及冰糖鋪在雞上，澆上餘下之黑糯米酒，隔水大火蒸25分鐘，關火，加蓋焗5分鐘，蒸雞汁留用。
3. 雞待冷後，斬件上碟，煮滾蒸雞之黑糯米酒汁，澆上雞件即成。

零失敗技巧
Successfully Cooking Skills

必須用黑糯米酒嗎？哪裏購買？

用黑糯米酒蒸雞，肉質嫩滑，入味可口，連骨都充滿酒的甜香味，怎能不試！一般超市有售。

Need to use black glutinous rice wine? Where to buy?

The chicken steamed in the wine is soft and gentle. It is so delicious that even the bone gives the nice smell of wine. How can we avoid trying! It can be bought at supermarkets.

用冰鮮雞烹調，鮮味略遜嗎？

當然，冰鮮雞的鮮味比鮮雞略遜，但配搭紅棗及黑糯米酒，依然美味！

Is frozen chicken inferior to fresh chicken in flavour?

Yes, of course. But it is still tasty by cooking with red dates and black glutinous rice wine.

為何雞蒸妥後，要關火再焗？

讓熱力令雞骨全熟，不帶血水，而且紅棗香、黑糯米酒的甜膩更能充份滲入雞肉！

Why leave the steamed chicken on stove after turning off the heat?

The remaining heat allows the bone to cook through with no blood oozing out, and the aromatic flavour of red dates and the sweetness of black glutinous rice wine to fully infuse into the chicken.

Ingredients (Serves 4)

1 premium chilled chicken
15 red dates
2 tbsp rock sugar, crushed
1 cup black glutinous rice wine
1 tsp salt
6 slices ginger

Method

1. Stone and rinse the red dates. Set aside. Rinse the chicken and drain.
2. Rub the chicken cavity with the salt. Put 1/2 portion of the red dates, 1/2 portion of the rock sugar, ginger and a little black glutinous rice wine into the chicken cavity. Place the rest of the red dates and sugar on the chicken. Sprinkle with the remaining wine. Steam over high heat for 25 minutes. Turn off heat. Leave the chicken with a lid on for 5 minutes. Reserve the sauce obtained from the steamed chicken.
3. When the chicken cools down, chop into pieces. Bring the sauce to the boil. Sprinkle on top of the chicken. Serve.

甜酒腐乳蒸雞

Steamed Chicken with Sweet Wine Fermented Beancurd

材料（4 人份量）

光雞 12 兩
甜酒豆腐乳 3 件（壓成茸）
甜酒豆腐乳豆粒 1 湯匙
葱絲 1 湯匙

調味料

生抽 2 茶匙
鹽 1/4 茶匙
薑汁 1 湯匙
麻油及胡椒粉各少許
生粉及油各半湯匙

做法

1. 光雞洗淨，抹乾水分，斬件，加入調味料、甜酒豆腐乳及豆粒拌勻，醃 15 分鐘，放入深碟內。
2. 燒滾水，放入雞件隔水蒸 12 分鐘，灑入葱絲再蒸半分鐘即成。

Ingredients (Serves 4)

450 g chicken
3 pieces sweet wine fermented beancurd, mashed
1 tbsp beans from sweet wine fermented beancurd
1 tbsp shredded spring onion

Seasoning

2 tsp light soy sauce
1/4 tsp salt
1 tbsp ginger juice
sesame oil
ground white pepper
1/2 tbsp caltrop starch
1/2 tbsp oil

Method

1. Rinse the chicken. Wipe it dry. Chop into pieces. Mix with the seasoning, sweet wine fermented beancurd, and beans. Rest for 15 minutes. Place in a deep dish.
2. Bring water to the boil. Steam the chicken over water for 12 minutes. Sprinkle the spring onion on top. Steam for 1/2 minute. Serve.

零失敗技巧
Successfully Cooking Skills

甜酒豆腐乳是甚麼？
這是台灣的調味料，以黃豆、糙米及米酒等發酵而成，色澤暗沉，發酵越久，香氣越濃，於台灣食品店有售。

What is sweet wine fermented beancurd?
Dark in colour, it is a Taiwanese condiment by fermenting soy bean, brown rice, rice wine and other ingredients. The longer it is preserved, the stronger aroma it gives. It can be bought at Taiwanese food shops.

為何加入甜酒豆腐乳的豆粒？
豆粒滲有濃濃的酒香味，伴雞件同蒸，非常惹味！

Why add the beans of the sweet wine fermented beancurd?
The beans have a strong aromatic flavour of wine. They will give the steamed chicken a sensational taste!

雞件醃 15 分鐘，足夠嗎？
絕對足夠！香味滲入雞肉。

Is it enough to marinate the chicken for 15 minutes?
Yes, absolutely. The aromatic flavour will infuse into the chicken.

煮人必學家常菜

Must-learn home-style recipes

編者
Forms Kitchen編輯委員會

Editor
Editorial Committee, Forms Kitchen

美術設計
羅穎思

Design
Venus Lo

排版
葉青

Typography
Rosemary Liu

出版者
香港鰂魚涌英皇道1065號
東達中心1305室
電話
傳真
電郵
網址

Publisher
Forms Kitchen
Room 1305, Eastern Centre, 1065 King's Road,
Quarry Bay, Hong Kong.
Tel: 2564 7511
Fax: 2565 5539
Email: info@wanlibk.com
Web Site: http://www.wanlibk.com
http://www.facebook.com/wanlibk

發行者
香港聯合書刊物流有限公司
香港新界大埔汀麗路36號
中華商務印刷大廈3字樓
電話
傳真
電郵

Distributor
SUP Publishing Logistics (HK) Ltd.
3/F., C&C Building, 36 Ting Lai Road,
Tai Po, N.T., Hong Kong
Tel: 2150 2100
Fax: 2407 3062
Email: info@suplogistics.com.hk

承印者
中華商務彩色印刷有限公司

Printer
C & C Offset Printing Co., Ltd.

出版日期
二零一九年一月第一次印刷

Publishing Date
First print in January 2019

ISBN 978-962-14-6962-5

鳴謝以下作者提供食譜（排名不分先後）：
黃美鳳、Feliz Chan、Winnie姐